Phytoremediation of Emerging Contaminants in Wetlands

Phytoremediation of Emerging Contaminants in Wetlands

Prabhat Kumar Rai

CRC Press
Taylor & Francis Group
Boca Raton London New York

CRC Press is an imprint of the
Taylor & Francis Group, an **informa** business

CRC Press
Taylor & Francis Group
6000 Broken Sound Parkway NW, Suite 300
Boca Raton, FL 33487-2742

© 2018 by Taylor & Francis Group, LLC
CRC Press is an imprint of Taylor & Francis Group, an Informa business

No claim to original U.S. Government works

Printed on acid-free paper

International Standard Book Number-13: 978-0-8153-8510-3 (Hardback)

Visit the Taylor & Francis Web site at
http://www.taylorandfrancis.com

and the CRC Press Web site at
http://www.crcpress.com

Contents

Preface

This book aims to interest readers in wetlands from an environmental perspective—apart from their aesthetic and scenic attributes. The past several decades have witnessed the perturbation of water and the environment from emerging contaminants, such as heavy metals, volatile organic carbons, pesticides, personal care products, pharmaceuticals, organics, NPs, and pathogenic microbes. These pollutants emanate from several human activities, and therefore should be of obvious interest and attraction to readers. Further, to understand the plant-based technologies for ameliorating water pollution from these emerging environmental contaminants, particularly heavy metals, a detailed analysis of phytoremediation, as well as its mechanisms and applied aspects, is presented. Moreover, a critical analysis of the strengths and weaknesses of phytoremediation technology imposes constraints that need to be addressed. This book details up-to-date progress in wetland science and phytoremediation all across the globe. It includes case studies of interest to people from both temperate and tropical countries. In one place, readers can find complete global research on wetlands, with case studies from both developing and developed countries across different continents. This design guide and methodology for water quality monitoring should be of interest to young students who are pursuing a future in water or wetland research. Finally, a concise description of the future prospects (genetic engineering, green chemistry, and nanoparticles [NPs]) is provided.

The book consists of nine chapters. Chapter 1 describes the diverse attributes related to water, wetlands, and the phytoremediation of heavy metals and other emerging environmental contaminants, like volatile organic carbons, pesticides, personal care products, pharmaceuticals, organics, and pathogenic microbes. Chapter 2, "Phytoremediation: Concept, Principles, Mechanisms, and Applications," provides a detailed analysis on environmental contaminants, particularly heavy metals, and their phytoremediation with wetland plants and macrophytes. Chapter 3, "Progress, Prospects, and Challenges of Phytoremediation with Wetland Plants," discusses the global developments in recent phytotechnologies for emerging environmental contaminants. Emphasis is given to heavy metals; however, other organic contaminants, like volatile organic carbons and pesticides, are also included. The strengths (advantages) and limitations of phytoremediation are also discussed. To this end, the fate of environmental contaminants like metals in wetland plant biomass and their eco-friendly solutions are discussed in this chapter. Chapter 4, "Natural and Constructed Wetlands in Phytoremediation: A Global Perspective with Case Studies of Tropical and Temperate Countries," deals with the natural and constructed wetlands of the world and their role in the phytoremediation of emerging contaminants, like organic pesticides and heavy metals. This chapter discusses how constructed wetlands are a natural alternative to technical methods of wastewater treatment. In this chapter, we also provide a brief account of the importance of wetland plants and their role in the phytoremediation process, particularly heavy metals. Moreover, in Chapter 4 we present case studies of treatment wetlands from different countries: the United States, European nations, and African and Asian countries. It is worth mentioning that these case studies for phytoremediation of emerging contaminants in wetlands exclude Ramsar sites, which we attempt to cover in Chapter 6. It is well known that in constructed wetlands, there may be several mechanisms of phytoremediation. Also, diverse types of constructed wetlands are discussed. Further, plant microbe interaction and the role of rhizospheric organisms are discussed. Moreover, the impact of environmental factors on the performance of constructed wetland treatment systems should also be taken into consideration. The role of hydraulic loading rate and hydraulic retention time and prospective features of constructed wetland treatment systems are also covered in this chapter. Chapter 5 concisely describes the methods and designer approach used in pollution science to assess the water quality of wetlands, which is inextricably linked to wetland management. Further, the water quality parameters of treatment wetlands noticeably affect the concentration as well as the phytoremediation mechanism of emerging contaminants, like heavy metals and organics. Chapter 6, "Global Ramsar Wetland Sites: A Case Study on Biodiversity Hotspots," describes global Ramsar sites with special reference to the wetland ecology of a Ramsar site of a global biodiversity hotspot. The Ramsar convention, which originated in Iran, identifies the wetlands of global tropical and temperate countries from different continents that need scientific attention. Thus, descriptions of global Ramsar sites provide their current global status. The biodiversity hotspots of Myers (1988), as published in the journal *Nature*, are conservation focus sites with an immense diversity of aquatic and terrestrial plants. Nevertheless, these global biodiversity hotspot sites are rarely investigated from a wetland plant and phytoremediation-based perspective, probably because of geographical constraints, among other ethnolinguistic constraints. To this end, biodiversity prospecting and screening of potent wetland plants in such biodiversity-rich regions may result in an eco-technological innovation for the treatment of global contaminants. Chapter 7 provides certain microcosm and field-scale investigations on the phytoremediation of metals with global wetland plants. Chapters 8 and 9 deal with the future perspective of phytoremediation with wetland plants, for example: the genetic engineering of wetland plants, the role of green chemistry, and NP generation. Further, recent concerns regarding the fate of NPs and the role of wetland plants in their phytoremediation are discussed in Chapter 9.

Wetland plants and NP interaction can transmogrify different environmental technologies inextricably linked to environmental remediation (phytoremediation, phytomining, phytosynthesis of NPs, and their mechanisms), besides having implications in the bioenergy sector. Thus, adequate coverage of green chemistry, the role of genetic engineering, and wetland plant–NP interactions makes this book different from other existing books.

REFERENCE

Myers N. 1988. Threatened biotas: "Hot-spots" tropical forests. *The environtalists* 8:178–208.

Author

Dr. Prabhat Kumar Rai is currently working as an assistant professor at Mizoram University, Aizawl, India. He has more than 10 years of teaching and research experience. He was awarded the Young Scientist Award in 2012 for his exemplary work. He has published 104 research papers in renowned journals and 10 books with publishers such as Elsevier and Nova Science, as well as contributed chapters in many other books.

Acknowledgments

I consider it a rare opportunity to thank Professor A.N. Rai, former director of National Assessment and Accreditation Council (NAAC) and vice chancellor of Mizoram University, Aizawl, India, as well as North-Eastern Hill University, Meghalaya, India, for his blessings, encouragement, and support. I am also indebted to Mrs. Urmila Rai for her affection and extending her love and blessings. I am extremely thankful to Professor R.P. Tiwari, Department of Geology, Mizoram University (current vice chancellor of Hari Singh Gaur University, Sagar, India), who always helped and encouraged me.

Pertaining to academic guidance, I am thankful to Professor J. Vymazal (for his remarkable contribution to wetland science and enriching the literature); J.N.B. Bell (Imperial College, London), my mentor for my last 10 years of research; Jason White (managing editor of *International Journal of Phytoremediation*); Terry Logan (editor in chief of *Critical Reviews in Environmental Science and Technology*); and Lee Newman (editor in chief of *International Journal of Phytoremediation*). I am also thankful to my teachers at Banaras Hindu University, Uttar Pradesh, India, especially Professor B.D. Tripathi, Professor J.S. Singh (FNA), Professor L.C. Rai (FNA), Professor Rajeev Raman (FNA), and Professor R.S. Upadhyay.

I am particularly thankful to Renu Upadhyay and Shikha Garg of CRC Press, a division of Taylor & Francis, for their continuous support and suggestions. To this end, I wish to thank all six reviewers for their suggestions, which helped me to upgrade this book to its present form.

I would like to thank my parents (Dr. Om Prakash Rai, ex-principal of CHS–Banaras Hindu University, and Mrs. Usha Rai) and family members (Prashant K. Rai [advocate], Pratibha Rai, and Dr. Ved Prakash Rai [scientist]). I also remember and admire the blessings of my grandparents, the late Braj Bihari Rai and Chandravati Rai. Further, thanks to my wife (Garima Rai), who has supported me all through this course, and her parents (Shri Vijai K. Rai and Smt. Asha Rai). I also pray for the well-being of souls who have recently left me, that they may rest in peace in their existing forms. Moreover, I would like to extend my love to little Pranjali Rai and Rachit Rai, as they bring a great deal of fortune with them.

Porobhot kumar Rai

Prabhat Kumar Rai
Department of Environmental Sciences
School of Earth Sciences and Natural Resource Management
Mizoram Central University, Tanhril, Aizawl, India

1 Water, Wetlands, and Phytoremediation of Emerging Environmental Contaminants

WATER: GLOBAL DISTRIBUTION AND WATER QUALITY (PHYSICOCHEMICAL) ATTRIBUTES

It is well known that water, an integral component of the environment, is responsible for various life processes, and hence persistence of life on earth. The salient features of water lie in its typical chemical constitution. Water is a substance composed of hydrogen and oxygen, and it exists in solid, liquid, and gaseous states. It is a colorless, tasteless, and odorless liquid at room temperature. Water exhibits very complex chemical and physical properties that are incompletely understood, although its formula seems simple. Its melting point is 0°C (32°F), and its boiling point is 100°C (212°F). Compared with analogous compounds, such as hydrogen sulfide and ammonia, it is much higher than the expected melting and boiling point. In its solid form (ice), water is less dense than when it is in liquid form, another unusual property.

Water is the major component of the human body, and if it is extracted from the body, one dies immediately. Water is essential for completion of the life cycle in animals and plants because it is needed at one stage or another of life. Water is essential for the survival of all living organisms. The important property of water is its ability to dissolve many substances; therefore, water is considered a universal solvent. The dissolved substances have a significant role in many mechanisms of living organisms. Life is also believed to have originated in the world's oceans, complicated solutions of water in which many substances are dissolved. Living organisms use aqueous solutions, for example, blood and digestive juices, as media for carrying out biological processes.

Water is the essence of life on earth and totally dominates the chemical composition of all organisms. The utility of water in biota is the fulcrum of biochemical metabolism, which rests on its unique physical and chemical properties. The characteristics of water regulate lake metabolism. The unique thermal density properties, high specific heat, and liquid–solid characteristics of water allow the formation of a stratified environment that controls the chemical and biotic properties of lakes to a marked degree. Water provides a tempered milieu in which extreme fluctuations in water's availability and temperature are ameliorated relative to the conditions facing aerial life. Coupled with a relatively high degree of viscosity, these characteristics

1

have enabled biota to develop a large number of adaptations that improve sustained productivity (Wetzel, 1983).

Of the total water reserves, the ocean contains 97%, ice caps and permanent glaciers 2.1%, and freshwater only 0.9%, which is found in atmospheric water vapor, rivers, lakes, ponds, groundwater, and soil moisture (Dugan, 1972) (Figure 1.1). Since less than 1% of the global water is available for sustaining the "global village" of human society, its proper conservation and management are prerequisites that are being adversely perturbed through diverse emerging contaminants.

As the earth's population grows and the demand for freshwater increases, water purification and recycling become increasingly important. Interestingly, the purity requirements of water for industrial use often exceed those for human consumption.

Although present in a larger amount, oceanic water is of little direct use. Freshwater resources have played an important role in the development and evolution of civilizations. This is because freshwater resources find their use in various human endeavors, like drinking, washing, transportation, and irrigation. Therefore, proper conservation and management are prerequisites, so that this small percentage of usable water can be sustained for various purposes that are directly linked to human survival.

Lakes and surface water reservoirs are the planet's most important freshwater resources and provide innumerable benefits. They are a source of water for domestic use, irrigation, and renewable energy in the form of hydropower, and are essential resources for industry. Lakes provide ecosystems for fish, thereby functioning as a source of essential protein, and for significant elements of the world's biological diversity. They have important social and economic benefits as a result of tourism and recreation, and are culturally and aesthetically important for people throughout the world.

Water has several unique physical and chemical properties that have influenced life as it has evolved. Indeed, the very concept of the earth's biosphere is dependent

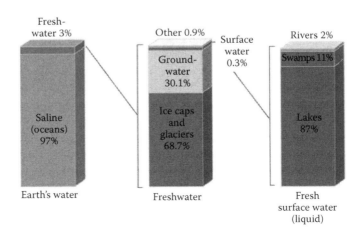

FIGURE 1.1   Quantitative account of global water distribution. (From Google.)

on the special physicochemical properties of water. These characteristics have significantly influenced the structure of inland aquatic ecosystems.

It is essential and important to check water before it is used for various purposes, for example, domestic, agricultural, and industrial purposes. Water must be tested with different physicochemical parameters. The selection of parameters for the testing of water is solely dependent on the purpose for which the water will be used and how high its quality and purity need to be. Water may contain different types of floating, dissolved, suspended, and microbiological as well as bacteriological impurities. Some physical tests should be performed for testing its physical characteristics, such as temperature, color, odor, pH, turbidity, transparency, and total solids, while chemical tests should be performed to determine its biological oxygen demand (BOD), dissolved oxygen (DO), acidity, alkalinity, chloride, hardness, nitrate, and phosphate. For obtaining more potable or pure water, it should be tested for its trace metals, heavy metal contents, and organic (i.e., pesticide) residues. It is obvious that drinking water should pass all these tests and contain the required amount of minerals. Only in developed countries are all these criteria strictly monitored. Due to the very low concentration of heavy metals and organic pesticide impurities present in water, highly sophisticated analytical instruments and well-trained manpower are needed. The following different physicochemical parameters are tested regularly to monitor the quality of water.

Temperature affects a number of physical, chemical, and biological processes in natural aquatic systems. The temperature regime of a water source is a function of the seasonal and diurnal ambient air temperatures and the morphometry and setting of the water source. Biologically, one of the most important effects of temperature is the decrease in oxygen solubility as the temperature increases. As a result, an increase in temperature can also increase the oxygen demand of biological organisms, such as aquatic plants and fish. In addition, chemical compounds tend to become more soluble as a response to higher temperatures. Temperature is controlled primarily by climatic conditions, but human activity can also influence it. Thermal or chemical pollution can adversely alter the distribution and species composition of aquatic communities.

At prevailing global temperatures, most inland water is in liquid form. Liquid water has special thermal features that minimize temperature fluctuations. First among these features is its high specific heat; that is, a relatively large amount of heat is required to raise the temperature of water. The quantity of heat required to convert water from a solid to a liquid state (latent heat of fusion) is high. This capacity to absorb heat has several important consequences for the biosphere, including the ability of inland water to moderate seasonal and diurnal temperature differences both within aquatic ecosystems and, to lesser extent, beyond them. Most of the heat input to inland waters is in the form of solar energy. The amount of heat that actually reaches inland waters at any time depends on several factors, including time of the day, season, latitude, altitude, and amount of cloud cover. A significant amount of solar radiation that reaches the water column surface is lost through reflection and backscattering. The remaining fractions enter the water column, where its energy rapidly diminishes with depth as it is absorbed and converted either to heat by physical processes or to chemical energy by the biological processes of photosynthesis.

In large lakes, most of the energy required by the biota is derived from this biological conversion. In other sorts of inland waters, however, a large proportion of the required energy by biological communities may come from emergent and nearby terrestrial vegetation. In any event, the amount and nature of solar energy entering inland waters is a principal determinant of the structure and function of the ecosystem.

One of the most significant properties of water is its function as a solvent. Water is considered a universal solvent. In this regard, it has an unrivalled capacity to dissolve in solution in an exceptionally wide range of substances, including electrolytes (salt, which dissociates into ions in aqueous solution), colloids (particulate matter small enough to remain suspended in solution), and nonelectrolytes (substances such as glucose that retain their molecular structure and do not dissociate into ions). A great variety of combinations of dissolved substances can occur in inland waters. The major inorganic solutes are the cations sodium, potassium, calcium, and magnesium and the anions chloride, sulfate, and bicarbonate or carbonate. When the concentration of all these ions (i.e., the salinity, or salt content) is less than 3 g/L (i.e., 3 g/kg or 3 parts per thousand), inland waters are conventionally regarded as fresh. In addition to these major ions, all inland waters contain smaller quantities of other ions, of which phosphate and nitrate essential plant nutrients are particularly significant and certain dissolved gases, especially oxygen, carbon dioxide, and nitrogen, whose solubilities are inversely correlated with temperature, altitude, and hydrogen ion concentration (pH), are of biological significance.

Physicochemical phenomena affect every body of inland water, creating unique relationships among and within the biotic and abiotic components of the ecosystem. Of particular interest are the pathways or biogeochemical cycles that are traveled by the chemical elements essential to life: nitrogen, phosphate, carbon, and a variety of micronutrients, such as iron, sulfur, and silica. The degree to which the output of a particular element balances the input within a given aquatic ecosystem varies according to the type of inland water involved. However, all essential elements follow pathways in inland waters that are numerous, complex, well-defined, and often interdependent on the other biogeochemical cycles. In fact, a defining characteristic of all inland aquatic ecosystems, including the simplest temporary bodies of highly saline water, is the occurrence of well-defined biogeochemical cycles (Wetzel, 1983).

Some of the most salient general physiocohemical features of inland waters are continuously varying attributes which in turn affect the hydrological behaviour and bioavailability of emerging contaminants.

Inland waters represent parts of the biosphere within which marked biological diversity, complex biogeochemical pathways, and energetic processes occur. Although from a geographic perspective inland waters represent only a small fraction of the biosphere, when appreciated from an ecological viewpoint, they are seen to be major contributors to biospheric diversity, structure, and function. This regulation of the entire physical and chemical dynamics of lakes and the resultant metabolism is governed to a great extent by differences in density. Density increases with an increase in concentrations of dissolved salts in an approximately linear fashion.

Water is a vital natural resource, which is essential for multiple purposes. It is an essential constituent of all animal and vegetable matter. It is also an essential ingredient of animal and plant life. Its uses may include drinking and other domestic uses, industrial cooling, power generation, agriculture, transportation, and waste disposal. At the present state of national development, the agricultural productivity in India, heavily dependent on rainfall and impacted by the droughts occurring in various parts of the country during the last decade, has given a series of jolts to the growth of the country's economy. The growing population, accelerating pace of industrialization, and intensification of agriculture and urbanization exert heavy pressure on our vast but limited water resources.

WATER POLLUTION: GLOBAL SOURCES

The Water (Prevention and Control of Pollution) Act of 1974 defines water pollution as

> such contamination of water and such alteration of the physical, chemical or biological properties of water or such discharge of any sewage or trade effluent or any other liquid, gaseous or solid substance into water (whether directly or indirectly) as may or is likely to, create a nuisance or render such water harmful or injurious to the public health or safety, or to domestic, commercial, industrial, agricultural or other legitimate uses, or to the life and health of animals or plants or of aquatic organisms.

With the increase in the age of the earth, clean water is becoming more precious as water continues to be polluted by several man-made activities, for example, rapid population growth, the alarming speed of industrialization and deforestation, urbanization, increasing living standards, and wide spheres of other human activities. Groundwater, surface water, rivers, sea, lakes, ponds, and so forth are finding it more and more difficult to escape from pollution. The term *water pollution* refers to anything causing change in the diversity of aquatic life. The presence of too much of an undesirable foreign substance in water is responsible for water pollution. Water pollution is one of the most serious problems faced by humans today, and a vital concern for humankind, being essential for humans, animals, and aquatic life. It is the universal enabling chemical that is capable of dissolving or carrying in suspension a variety of toxic materials from mainly the heavy flux of sewage, industrial effluents, and domestic and agricultural waste. That is why it is of special interest to study water pollution.

SOURCES OF GLOBAL WATER POLLUTION

The sources of contamination of water are as follows:

1. Sewage and domestic wastes
2. Industrial effluents
3. Agricultural discharges
4. Pesticides and fertilizers
5. Soap and detergents
6. Thermal pollution

Sewage and Domestic Wastes

About 75% of water pollution is caused by sewage and domestic wastes. If domestic waste and sewage are not properly handled after they are produced and are directly discharged into water bodies, the water becomes polluted. Domestic sewage contains decomposable organic matter that exerts an oxygen demand, and sewage contains oxidizable and fermentable matter that causes depletion of the DO level in water bodies.

Industrial Effluents

Industrial effluents are discharged into water bodies containing toxic chemicals, phenols, aldehydes, ketones, cyanides, metallic wastes, plasticizers, toxic acids, oil and grease, dyes, suspended solids, radioactive wastes, and so forth. The principal types of industries that contribute to water pollution are chemical and pharmaceutical, steel, coal, soap and detergents, paper and pulp, distilleries, tanneries, and foods processing. These effluents, when discharged through sewage systems, poison the biological purification mechanism of sewage treatment and pose several pollution problems.

Agricultural Discharge

Plant nutrients, pesticides, insecticides, herbicides, fertilizers, and plant and animal debris are reported to cause heavy pollution to water sources. Nowadays, fertilizers containing phosphates and nitrates are added to soil; some of these are washed off through rainfall, irrigation, and drainage into water bodies, thereby severely disturbing the aquatic system. Organic wastes increase the BOD of the receiving water body. Some pesticides are nonbiodegradable and, when sprayed, remain in the soil for a long time before being carried into water bodies during rainfall.

Soap and Detergents

Soap forms oleic acid and fatty acid when in contact with water. Thus the acidity of water increases, which perturbs the aquatic life. Detergents used as cleaning agents contain several pollutants that severely affect water bodies. They contain surface activity agents and contribute to the phosphates of sodium, silicates, sulfates, and several other salt builders in water. Wastewater contaminated with detergents carries a huge cap of foam, which is an anaesthetic for all purposes. Since detergents are composed of complex phosphates, they increase the concentration of phosphates in water, making it poisonous and causing eutrophication problems.

Thermal Pollution

The discharge of pollutants with unutilized heat from nuclear and thermal power adversely affects the aquatic environment. Apart from the electric power plant, various other industries with cooling systems contribute to the thermal loading of water bodies. These pollutants increase the temperature of water and decrease the DO value, thus making conditions unsuitable for aquatic life.

IMPORTANCE OF GLOBAL WETLAND PLANTS

Macrophytes and global wetland plants play a major role in the structural and functional aspects of aquatic ecosystems by altering water movement regimes, providing shelter to fish and aquatic invertebrates, serving as a food source, and altering water quality by regulating oxygen balance and nutrient cycles and accumulating heavy metals (Rai, 2007a,b, 2008b, 2011; Sood et al., 2012). Macrophytes have the natural ability to take up, accumulate, or degrade organic and inorganic substances (Lasat, 2000; McIntyre, 2003) and heavy metals through the process of bioaccumulation (Tiwari et al., 2007; Rai, 2010a–d; Rai et al., 2010). On the other hand, macrophytes can bioaccumulate, biomagnificate, or biotransfer certain metals to concentrations high enough to bring about harmful effects (Opuene et al., 2008). Some macrophyte species are adapted to grow in areas of higher metal concentrations (Chatterjee et al., 2011), some of which have the indispensable property of metal tolerance (Singh et al., 2003; Chatterjee et al., 2004; Bertrand and Poirier, 2005). Various species show different behaviors regarding their ability to accumulate elements in roots, stems, and/or leaves (Kumar et al., 1995). Therefore, it is necessary and useful to identify the plant organ that absorbs the greatest amount of trace elements (St-Cyr and Campbell, 1994; Baldantoni et al., 2004).

In wetland aquatic ecosystems where pollutants are discontinuous and quickly diluted, analyses of plant components provide time-integrated information about the quality of the ecosystem (Baldantoni et al., 2005). Conventional technologies, such as electrolysis, reverse osmosis, and ion exchange, were used for cleaning heavy metal (Rai, 2009a). The use of such macrophyte species in the remediation of water contaminated with heavy metals is a promising cost-effective alternative to the more established treatment methods (Chatterjee et al., 2011).

WETLANDS, PLANTS, AND PHYTOREMEDIATION OF EMERGING CONTAMINANTS

Water pollution of wetlands from diverse emerging contaminants is a global problem still in search of an affordable (cost-effective) and eco-technological solution. Pollutants having no prescribed regulatory limits are usually referred to as emerging micropollutants; these consist of endocrine disruptors and pharmaceuticals and personal care products (PPCPs) (Jiang et al., 2013). The past couple of decades have witnessed diverse chemicals and emerging contaminants, like heavy metals, volatile organic carbons, pesticides, personal care products (PCPs), pharmaceuticals, organics, and pathogenic microbes, perturb the water and environment.

In view of several disadvantages and side effects of chemical and other conventional technologies, global research has started on a quest for alternative environmentally friendly phytotechnologies. Being at the interfaces of water, air, and soil, plants can be a better tool to elucidate the mechanisms of removal of emerging contaminants through phytoremediation. To this end, phytoremediation of emerging contaminants with wetland plants or macrophytes is an eco-friendly, aesthetically pleasing, cost-effective, solar-driven, passive technique for environmental management. Thus, phytoremediation has become an increasingly recognized pathway for the removal of emerging contaminants from the environment. (Dushenkov and Kapulnik, 2000; Farraji et al., 2016)

Wetlands (both natural and constructed) are habitats for a wide variety of plant and animal life, especially waterbirds. Constructed wetlands (composed of wetland plants and microbes) are a cost-effective, eco-sustainable tool for the phytoremediation of diverse emerging contaminants (Vymazal and Brezinova, 2015). The natural ability of wetland ecosystems to improve water quality has been recognized since the 1970s (Knight et al., 1999).

Apart from harboring birds, wetlands are also a nursery for several species of fish and shellfish, and a variety of organisms. Coastal wetlands, especially being an ecotone between the sea and freshwater, and/or freshwater and terrestrial habitats, have high species diversity. Ecologically, wetlands also perform important functions. They regulate the water regime, act as natural filters, and display amazing nutrient dynamics. As an ecosystem, wetlands are useful in recovering and cycling nutrients, releasing excess nitrogen, deactivating phosphates, treating wastewater, and removing toxins, chemicals, organic pesticides, and heavy metals through absorption by plants. Mangrove wetlands in coastal regions act as a buffer against devastating storms. Therefore, wetlands help in mitigating floods, recharging aquifers, and reducing surface runoff and the consequent erosion. In a nutshell, wetlands improve the quality of water and keep the local weather moderate. Wetlands in the urban periphery are natural receptacles for wastewater and can effectively harness the nutrients available in waste through fisheries and agriculture. Wetlands' shallow depth and full exposure of sunlight stimulate chemical processes, while larger input owing to higher productivity stimulates bacterial processes. These unique characteristics of wetlands promote various biogeochemical processes that are responsible for the extraordinary capacity of wetlands to retain emerging environmental contaminants, like heavy metals and organic pollutants, from a diverse range of industrial and agricultural effluents.

Wetlands possess the complex mechanism to remove the emerging contaminants through sorption (adsorption or absorption). Interestingly, plants are integral components of both natural and constructed wetlands in the remediation of emerging contaminants (Kadlec et al., 2010). In this context, hyperaccumulator plants, specifically wetland plants, tend to revitalize the environment from various emerging contaminants through phytoremediation, an eco-friendly phytotechnology. To elucidate the fate of emerging contaminants in aboveground and belowground plant parts of wetland plants, mechanisms linked with phytoremediation and biochemical changes are integral to phytoremediation technology for their increased efficiency. To this end, genetic engineering, omics (proteomics, genomics, metabolomics, etc.), and technological innovation as novel instruments made remarkable advances in boosting the efficiency of phytoremediation technology.

Concept, principles, and mechanisms of phytoremediation gradually advanced in due course, and a timely update is covered in this book in an interesting as well as systematic way. The different actions of plants and their associated rhizosphere bacteria on contaminants include phytoextraction, rhizofiltration, phytostabilization, phytodegradation, rhizodegradation, and phytovolatilization. Microbial associations with wetland plants are inextricably linked with phytoremediation success, as demonstrated in rooted wetland plants *Pistia stratiots* and *Eichhornia crassipes*, whose exudates supported bacterial strains eventually assisting in contaminant removal. Also, it is worth mentioning that emerging contaminant interactions in dicot wetland plants can lead to plaque formation near roots, while in the case of monocots, there is formation of phytosiderophores.

Pertaining to mechanisms involved in phytoremediation, the phytostabilization process tends to reduce the mobility and bioavailability of contaminants through their precipitation inside roots and the rhizosphere, thus protecting the aboveground plant parts from toxic contaminants. However, this technique is applicable only to rooted wetland plants like *Typha domingensis* and *Phragmites australis*. Further, phytotransformation and phytodegradation is used to degrade or immobilize contaminants in roots and shoots of plants with the help of enzymes or their cofactors. Another mechanism, phytoextraction, is the most vital technique for contaminated soil and water; however, it should not be done with agriculture crops in view of adverse human health impacts, in addition to contamination in the food chain. In phytoextraction, both aboveground (shoot) and belowground (root) parts of plants are involved and tend to transport and concentrate a diverse range of contaminants. Similarly, rhizofiltration does the same as phytoextraction, but with the involvement of the belowground portion, that is, root only. In the current scenario, the mechanisms involved in the phytoremediation of emerging contaminants are elucidated at the molecular level; however, scanty research exists in unraveling the phytoremediation pathway of nanoparticles (NPs) and magnetic NPs present in dust and particulate matter in Myers global biodiversity hotspots (Pandey et al., 2005; Rai, 2009b, 2012a, 2013, 2015a–d, 2016a,b; Rai et al., 2013, 2014, 2018; Rai and Panda, 2014; Rai and Singh, 2015; Rai and Chutia, 2016a; Singh and Rai, 2016). Phytoremediation in wetland plants or macrophytes involves plant–soil–microbe interactions, cell wall binding, cytoplasmic chelators, symplastic loading, ion exchange, and so forth, which are discussed in this book. Moreover, both the pros and cons (or strengths and weaknesses) associated with wetland plants in phytoremediation are discussed in Chapter 3. Detailed analyses of phytoremediation with wetland plants and linking its important attributes are done in this project.

The fate of emerging contaminants, especially heavy metals, in plant biomass is an issue that is rarely addressed and has been raised as the most limiting part of phytoremediation by several stakeholders and a group of researchers. Environmental contaminants are cleaned by hyperaccumulator wetland plants through phytoremediation by loading them in aboveground and belowground biomass. Nevertheless, there exists an environmental concern after phytoremediation, attributed to the probability of leakage of these contaminants into the environment again; thus, concerns over the ultimate fate and safe disposal of these contaminant-loaded biomasses puzzle phytotechnologists, and possible solutions to this problem are addressed in this book. Also, the invasive nature of certain wetland plants exacerbates the problem. Nevertheless, biomass and biosystems of invasive plants aid in environmental amelioration and the cost-effective sustainable management of emerging contaminants. On the other hand, other conventional management of invasive plant removal is costly and not eco-sustainable in nature (Rai, 2012a, 2013, 2015a–d, 2016a; Rai and Chutia, 2016b; Rai and Singh, 2016; Lee et al., 2017). To this end, I also suggest a couple of ways to sustainably manage the biomass of potent phytoremediation tools, that is, to phytosynthesize NPs (specifically metal NPs) and to couple their biomass for green renewable energy production. Also, to enhance the phytoremediation efficiency and completely eliminate contaminants from the environment, a *"designer" plant approach* can be of paramount importance, in addition to bioenergy and additional eco-friendly applications. It is well known that microbial association with plants is pertinent for the success of phytoremediation. In the designer plant approach, degrading microbes are integrated into the biosystem before transferring the plants to polluted sites to permit degradation of contaminants within the biomass.

PHYTOREMEDIATION OF EMERGING CONTAMINANTS (ORGANICS, PAHS, AND HEAVY METALS)

Most research performed worldwide has demonstrated BOD, chemical oxygen demand (COD), suspended solids, and nutrient removal; however, little research has been conducted on the removal of emerging environmental contaminants (volatile organic carbons, pesticides, PPCPs, organics, and pathogenic microbes) and heavy metals with global wetland plants. Further, recent concern on fate of NPs and the role of wetland plants in their phytoremediation is completely lacking elsewhere. Moreover, a designed approach, as well as a detailed methodology, involved in water quality monitoring and eco-management through phytoremediation is covered in this book. Additionally, as a green tool for emerging contaminants, plants possess remarkable potential for the phytoremediation of NPs, which have not been thoroughly evaluated.

Heavy metal pollution is a well-identified global problem that will be discussed separately in forthcoming chapters (specifically Chapter 3). Further, organic contaminants pose a serious threat to the ecosystem and human health, due to their toxicity, high frequency of occurrence, and/or persistence in the environment. Among compounds transported across plant membranes, volatile organics may be released through the leaves to the atmosphere via transpiration, while nonvolatile organics undergo sequestration or degradation inside the plant through phytoextraction and phytodegradation. The phytovolatilazation process can degrade organic contaminants like acetone, phenol, chlorinated benzenes, chlorinated ethenes, and BTEX (benzene, toluene, ethylbenzene, xylene) (Imfeld et al., 2009; Herath and Vithanage, 2015). Further, the Henry coefficient, high water solubility, and strong recalcitrance determine the removal of methyl t-butyl ether (MTBE) in constructed wetlands under anaerobic conditions (Imfeld et al., 2009). Moreover, organics can be degraded within the rhizosphere, through microbial activity (phytostimulation or rhizodegradation) (Wallace and Kadlec, 2005; Campos et al., 2008; Gerhardt et al., 2009; Afzal et al., 2014; Etim, 2012; Dordio and Carvalho, 2013; Lv et al., 2016a,b, 2017). Emerging contaminants, like polybrominated diphenyl ethers (PBDEs), were removed by phytoremediation with the common reed, that is, *Phragmites australis*, in conjunction with TiO_2 photocatalysis (Chow et al., 2017). The unique but popular wetland plant waterweed or water hyacinth (*Eichhornia crassipes*) possesses the potential to detoxify emerging contaminants, specifically organics like pentachlorophenol, perchloroethylene (PCE), and trichloroethylene (TCE) (Roy and Hanninen, 1994; Susarla et al., 2002). In constructed wetlands, a water quality analysis of organic contaminants may assist in designing treatment wetlands (Imfeld et al., 2009). Further, TCE has been remediated efficiently by phytoremediation of wetland species, including cattails (*Typha latifolia*), cottonwoods (*Populus deltoides*), and hybrid poplars (*Populus trichocarpa* × *P. deltoides*) (Gordon et al., 1998;

Bankston et al., 2002; Williams, 2002; Amon et al., 2007; Herath and Vithanage, 2015). Interestingly, *Typha latifolia* and *P. deltoids* removed TCE in the range of 73%–96%. To this end, hybrid poplars (*P. trichocarpa* × *P. deltoides*) resulted in phytoremediation of this organic emerging contaminant (TCE) up to 95% (Gordon et al., 1998). Also, other deleterious organic emerging contaminants (PCE) remediated the subsurface-constructed wetland (Amon et al., 2007).

For emerging contaminants of concern, like pesticides, wetland plants, such as *Ceratophyllum demersum*, *Elodea canadensis*, and *Lemna minor*, assisted in the rapid biotransformation of metolachlor and atrazine from herbicide-contaminated wastewaters or effluents (Williams, 2002). Another potent wetland plant, Bulrush (*Scirpus validus*), phytoremediated simazine and metolachlor from wastewaters or effluents in subsurface flow-constructed wetland (Stearman, 2003). Further studies also certified use of constructed wetland in remediating an emerging contaminant of particular concern i.e., chlorpyrifos in Cape Town city, South Africa. (Moore et al., 2002).

Moreover, oil or hydrocarbon contamination hampers salt marshes or natural wetland ecosystem sustainability. In this context, *Juncus roemerianus* proved to be an efficient phytoremediator for hydrocarbon, petroleum, or diesel-contaminated wetlands (Lin and Mendelssohn, 2009, 2012; Ribeiro et al., 2014; Syranidou et al., 2017a). Other wetland plants, like *Juncus effusus*, *Iris pseudacorus*, *Phalaris arundinacea*, and a grass mix (comprising *Festuca arundinacea*, *Festuca rubra*, and *Lolium perenne*), were used for the phytoremediation of polycyclic aromatic hydrocarbons (PAHs) (Leroy et al., 2015; Syranidou et al., 2017a). Explosives containing the emerging contaminant of concern, like 2,4,6-trinitrotoluene (TNT) transformation by aquatic plants, were observed using common wetland species *Myriophyllum spicatum* and *Myriophyllum aquaticum* (Williams, 2002) and in constructed wetland remediation occurred for TNT and Research Department Explosive (RDX) (Medina and McCutcheon, 1996; Best et al., 1999). The constructed wetland technology in its various forms (also in conjunction with wetland plants like *Typha* spp. and *Scirpus lacustris*) is used with a success rate of 95% for the removal of emerging contaminants like PAHs, which have been declared a priority pollutant by the U.S. Environmental Protection Agency (EPA) as well (Rezg et al., 2014; Terzakis et al., 2008; Fountoulakis et al., 2009; Haritash and Kaushik, 2009).

Bisphenol A (BPA) (2,2-bis[4-hydroxyphenyl]-propane) is a synthetic compound used in the production of polymers, vinyl chloride, thermal paper, and polyacrylates (Michałowicz, 2014; Syranidou et al., 2017b). BPA is globally produced (3 million tons each year), and wastewater treatment plants treating industrial effluents, sewage sludge, and waste landfill leachates are considered the main sources of BPA release into the environment (Syranidou et al., 2017b). Pertaining to the remediation of emerging contaminants, the wetland plant *Juncus acutus*, a helophyte, was found to be extremely useful. BPA is extremely deleterious to human health by causing several chronic diseases, like cardiovascular diseases, reproductive disorders, chronic kidney diseases, birth defects and development disorders, respiratory diseases, cancers, and autoimmune diseases. Deleterious emerging contaminants in the organic pollutants category, for example, BPA, efficiently remediated with this wetland plant (*J. acutus*) in association with the endophytic bacterial community (Syranidou et al., 2017b). It is worth mentioning that helophytes, halotolerant helophytes, and halophytes that have a higher tolerance to environmental contaminants are particularly useful for phytoremediation technology (Manousaki and Kalogerakis, 2011; Syranidou et al., 2017a). Helophytes are capable of pumping atmospheric oxygen to root tissues and releasing it into the rhizosphere, thus controlling microbial respiration by creating gradients of redox conditions (Stottmeister et al., 2003; Syranidou et al., 2017a). Interestingly, these endophytic microbes used BPA and/or two antibiotics (ciprofloxacin and sulfamethoxazole) as a carbon source to fulfill their nutritional requirements. In view of this potential, *J. acutus* is a potent plant that can be incorporated into constructed wetlands for its metal as well as organic contaminant resistance in soil and groundwater (overall emerging contaminant phytoremediation efficiency) (Syranidou et al., 2017a), and concomitantly, its efficiency is remarkably augmented by association of the microbial consortium. The association of such microbes greatly facilitated the remediation of carbamazepine and other psychotropic drugs through common potent wetland *Phragmites australis* (Sauvêtre and Schröder, 2015). Furthermore, synergistic association of *J. acutus* and endophytic microbes reduced the more toxic form of heavy metal Cr(VI) to the less toxic one, Cr(III) (Bais et al., 2006; Phillips et al., 2012; Syranidou et al., 2017b). The functionality of microbial communities in wetland plants *Juncus effusus*, *Typha latifolia*, *Berula erecta*, *Phragmites australis*, or *Iris pseudacorus* boosts the pesticide and organics phytoremediation (Lv et al., 2017). Table 1.1 lists the phytoremediation of emerging contaminants with wetland plants.

Emerging contaminants in the total environment, usually found at trace concentrations, that is, between nanograms per liter and micrograms per liter, or even lower, are considered micropollutants. In a recent report, 41 organic priority pollutants, 8 other substances with environmental quality standards (EQSs) listed in Directive 2013/39/EU, and 17 emerging contaminants of the Watch List of Decision 2015/495/EU, are reviewed for their removal in the context of constructed wetlands (Gortio et al., 2017). For these environmental micropollutants, conventional methods and sewage treatment plants were not found effective, which paved the way to constructed wetlands (Gortio et al., 2017).

According to one recent study, emerging contaminants, specifically hydrophobic organic contaminants, for example, dechlorane plus isomers, phyto-accumulated inside leaves of emergent *Typha angustifolia* existing in a large-scale-constructed wetland system (Wang and Kelly, 2017). Likewise, emerging contaminants in the total environment like, PBDEs, are not effectively treated with sewage treatment plants, and to this end, a combined photocatalysis (TiO_2 and visible light) and constructed wetland system (planted with *Oryza sativa* [rice cultivar Hefengzhan] and *Phragmites australis* [common reed]) were found to be successful with application of phytoremediation (Chow et al., 2017).

PHYTOREMEDIATION OF PPCPs (EMERGING CONTAMINANTS OF CONCERN) WITH WETLAND PLANTS

Interestingly, a specific category under emerging contaminants comprising endocrine-disrupting chemicals, pesticides, and PPCPs is well known for its efficiency to perturb the environment. Further, PPCPs usually derive their origin from industrial,

TABLE 1.1
Global Mesocosms or Pilot-Scale Phytoremediation Project

Wetland Plants	Emerging Contaminants	Global Sites, Both Tropical and Temperate	Remarks (Reference)
Pteris vittata (fern)	Arsenic (metalloid)	United States	1900 L/day, total 60,000 L drinking water (Eiless et al., 2005)
Phragmites australis	Organic contaminants	Germany	Site corresponds to contaminated aquifers (Braeckevelt et al., 2008)
Phragmites australis	Heavy metals	Taiwan	River water in three pilot-constructed wetlands with 180 × 50 × 50 cm diameters (Yeh et al., 2009)
Alternanthera philoxeroides	Effluents from dyes	India	96 h degradation time set for sulfonated textile dye (Rane et al., 2015)
Paulownia tomentosa	Heavy metals	Italy	Duration set for 30 days (Doumett et al., 2010)
Phragmites australis; Typha latifolia	Organic contaminants	United States	Pilot-constructed wetland with 35 m² (Ranieri et al., 2013)
Paspalum sp. + Tamarix gallica	Heavy metals	Italy	Phytoremediation for dredged marine sediment (Doni et al., 2015)
Phragmites australis	Organic contaminants	Germany	One year for contaminated aquifers; 2 years treatment + ferric oxide and charcoal augmentation (Braeckevelt et al., 2011; Seeger and Kuschk, 2011)
Typha latifolia	Metals from coal/lignite	Germany	4 years' subsurface flow-constructed wetland (Kuschk et al., 2003)
Phragmites australis	Organic contaminants mixed with nutrients	United Kingdom	4.8 m³ daily on a piggery farm-constructed wetland (Sun et al., 2006)
Typha latifolia, Phragmites australis, Iris pseudacorus, and Juncus effuses	Organic pesticides	Denmark	Imazalil and tebuconazole in plant tissue were relatively low (2.8%–14.4%) (Lv et al., 2016b)

(Continued)

TABLE 1.1 (CONTINUED)
Global Mesocosms or Pilot-Scale Phytoremediation Project

Wetland Plants	Emerging Contaminants	Global Sites, Both Tropical and Temperate	Remarks (Reference)
Typha domingensis, Ludwigia sp., Pistia stratiotes, and *Paspalum vaginatum*	Heavy metals (Cd, As, Hg, Cu, and Pb)	Coastal wetland in Ghana	*Typha domingensis, Ludwigia sp.,* and *Paspalum vaginatum,* respectively, had the highest accumulation capacities for Cd, As, and Hg, but the floating aquatic plant *Pistia stratiotes* appeared to be a better accumulator of Cd and As (Gbogbo and Otoo, 2015)
Phragmites communis, Salix viminalis, and *Populus canadensis*	Heavy metals (Ni, Sr, V, Zn, Cd, Cu, and Pb)	Constructed wetlands in Poland	*Salix* wetland system was most effective in purification overall, while *Populus* wetland system was most effective for Cu and Ni (Samecka-Cymerman et al., 2004)
Limnocharis flava, Thalia geniculata, and *Typha latifolia*	Emerging contaminants: heavy metals like Fe, Cu, Zn, Pb, and Hg	In a constructed wetland of Ghana	Phytoremediation of water making it suitable for irrigation in agriculture fields of Ghana (Anning et al., 2013)
Scirpus grossus (microbial consortium of three rhizobacteria strains: *Bacillus aquimaris, Bacillus anthracis,* and *Bacillus cereus*)	Emerging contaminant: hydrocarbon (PAH)	In a constructed wetland of Malaysia	Efficient phytoremediation of total petroleum hydrocarbon (TPH) effluent in HF-type-constructed wetlands with *Scirpus grossus* assisted with microbial consortium (Al-Baldawi et al., 2017)
Vetiver (*Chrysopogon zizanioides* L.)	Emerging contaminant: BOD and COD (from palm oil mill)	In a constructed wetland (floating vetiver system) of Malaysia	Modeling approach to treat the contaminants from palm oil mill effluent using response surface methodology (RSM) for optimization (Darajeh et al., 2016)
Typha angustifolia	Emerging contaminant: hydrophobic organic contaminants, e.g., dechlorane + isomers	In large-scale-constructed wetland system	Emerging contaminants like dechlorane + isomers phyto-accumulated inside leaves of emergent *Typha angustifolia* (Wang and Kelly, 2017)
Oryza sativa and *Phragmites australis* + combined photocatalysis (TiO_2 and visible light)	Emerging contaminant: PBDEs	In large-scale-constructed wetland system	PBDE phytoremediation (Chow et al., 2017)

municipal, and agricultural effluents. Interestingly, PPCPs are integral components of "nutraceuticals," fragrances, sunscreen agents, and widely used drugs, such as analgesics and anti-inflammatories, antidiabetics, antiestrogenics, antiprotozoals, antiseptics, lipid regulators, diuretics, medications for treating erectile dysfunction and pulmonary arterial hypertension, psychiatric drugs and antidepressants, psychostimulants, veterinary and human antibiotics, and beta-blockers. They are also used in x-ray and contrast media, cosmetics and PCPs, surfactants, and phytosanitary products which eventually act as emerging contaminants in environmental context. Unfortunately, approximately 15,000 tons of antibiotics are released annually by European countries alone. To this end, the total global consumption of ca. 4,000 antibiotics is estimated to release ca. 100,000–200,000 tons of PPCPs into the ecosystem (Hyland et al., 2015; Sui et al., 2015; Bartrons and Peñuelas, 2017).

In addition to perturbing environmental health and ecologic change, dietary intake of these PPCP-contaminated plants may also pose a severe threat to human health, but in the present scenario, little is known about the fate of PPCPs in plants and their effect on or risk to the ecosystem. PPCPs reach plants predominantly from the use of reclaimed wastewater for irrigation, the application of biosolids (treated sewage sludge) and manure for the fertilization of agricultural soils, and deposition from volatilized compounds (Daughton and Ternes, 1999; Hyland et al., 2015; Sui et al., 2015; Bartrons and Peñuelas, 2017).

Wetland and other aquatic ecosystems are particularly sensitive to PPCPs. A category of PPCPs, chemosensitizers, that is, those chemicals that inhibit multixenobiotic transporters, may play a key role in potentiating the effects of PPCPs. Prevention of direct discharge of untreated sewage to the environment would have the greatest impact on reducing the discharge of less persistent PPCPs (Daughton and Ternes, 1999). Moreover, attention may be warranted to ensure that the degradation of PPCPs to innocuous products in waste treatment plants is maximized (Daughton and Ternes, 1999).

Wetland plants may be tolerant to PPCPs, as demonstrated by *Lemna gibba* L. exposed to ibuprofen, which was found to exhibit no phytotoxic effects through growth and physiological parameters (Bacci et al., 2017). Henceforth, the phytoremediation of PPCPs through wetland plants is the most environmentally sustainable way of managing these emerging contaminants.

Juncus species possess the potential to sequester metals inside their aboveground as well as belowground biomass (Syranidou et al., 2017a), and some species of *Juncas* (like *J. phaeoce-phalus* and *J. patens*), in conjunction with another wetland plant, *Hydrocotyle* spp., removed organic contaminants like diazinon, dichlorodiphenyl-trichloroethane (DDT), total chlordanes, and total organochlorine pesticides (Carvalho et al., 2011; Lv et al., 2016a; Syranidou et al., 2017a). To this end, 16 PPCPs, such as ibuprofen, salicylic acid, caffeine, and triclosan, were removed with reasonable success using *Juncas effuses* (Reyes-Contreras et al., 2011). PPCPs like triclosan, methyltriclosan, and triclocarban were phytoremediated inside the biomass of three constructed wetland plants: *Typha latifolia*, *Pontederia cordata*, and *Sagittaria graminea* (Zarate et al., 2012).

Plants from wetlands and agro-ecosystems interact with PPCPs (usually considered pseudopersistent in nature), an emerging contaminant of concern, in a complex and intricate manner. Biomedicines have made remarkable advances in the recent past (Grassi et al., 2013; Carvalho et al., 2017). Nevertheless, the active ingredients in pharmaceuticals are transformed only partially in the bodies of living organisms and humans, and thus are excreted as a mixture of metabolites and bio-active forms into sewage and municipal treatment systems (Kim and Aga, 2007; Du and Liu, 2012; Carvalho et al., 2014; Rai et al., 2018). In this context, the emerging contaminants of concern are antibiotics, hormones, analgesics and anti-inflammatory drugs, chemical compounds used for disinfection and cleaning, beta-blockers, cancer therapeutics, contraceptives, hormones, and endocrine-disrupting compounds. Unfortunately, sewage treatment plants are unable to completely degrade these PPCPs or their partially degraded intermediates, and hence most rivers, lakes, and coastal waters, as well as the groundwater and some drinking water in populated areas, become contaminated (Ternes, 1998; Daneshvar et al., 2012; Carvalho et al., 2014). Further, agriculture systems are also recipients of untreated PPCPs from irrigation water from treatment plants that may contaminate the plants, which constitute the major portion of human dietary requirements (Rai, 2008a, 2012b; Rai and Tripathi, 2008). Therefore, serious human health risks can occur with the spread of this emerging contaminant of concern. Moreover, soil amendments with sewage sludge also exacerbate the problem, resulting in soil contamination. Also, the application of animal manure using veterinary antibiotics spreads them on agricultural land and contributes to their release into the environment (Kumar et al., 2012; Carvalho et al., 2014). Research of paramount importance was done using *Phragmites australis* for the phytoremediation of veterinary drugs, and the removal of 94% and 75% of enrofloxacin and tetracycline, respectively, was observed from spiked wastewater. Likewise, *Chrysopogon zizanioides* (vetiver grass) remediated tetracycline and accumulated this antibiotic mainly in aboveground biomass (Dutta et al., 2013).

Role of Constructed Wetlands in the Phytoremediation of PPCPs as Emerging Contaminants of Concern

Constructed wetland systems have the potential to treat water contaminated with PPCPs (Matamoros and Bayona, 2006; Verlicchi et al., 2013; Carvalho et al., 2014). Gerhardt et al. (2009) reviewed the phytoremediation and rhizoremediation of organic soil contaminants—both the potential and challenges. Carvalho et al. (2014) provided a detailed list of crop plants used in constructed wetlands and their interaction with PPCPs.

Emerging contaminants categorized as PPCPs, like sulfadimethoxine and flumequine from wastewaters, were investigated with *Azolla filiculoides* (Forni et al., 2001). With visualization or observation of some phytotoxicity, *A. filiculoides* was able to degrade the sulfadimethoxine and flumequine. To this end, the sulfadimethoxine uptake and degradation rates increased with concentrations in the culture medium (Forni et al., 2002). Other wetland plants, *Lythrum salicaria* and *Lemna minor*, were investigated with the emerging contaminant flumequine, and phytotoxic responses were observed in the form of root and shoot degradation, along with chlorophyll b decline in *L. minor* (Migliore et al., 2000; Cascone et al., 2004). However, both plants survived and demonstrated their tolerance behavior for 3–5 weeks under the effect of this pharmaceutical. Further, *L. minor* responded to other pharmaceuticals, like erythromycin (which resulted in growth inhibition by 20% at 1 mg/L) and tetracycline (which surprisingly promoted growth by 26% at 10 µg/L). Another common PPCP in drugs, ibuprofen, resulted in growth inhibition by 25% at 1 mg/L. In this context, exposure to the three pharmaceuticals resulted in the production of the stress hormone abscisic acid as a mechanism to forego the stress. Interestingly, erythromycin and tetracycline were more effective in promoting abscisic acid synthesis than ibuprofen (Pomati et al., 2004). Moreover, the duckweed *Lemna gibba* also demonstrated interaction with sulfamethoxazole, resulting in a phytotoxic response toward this PPCP (Brain et al., 2008).

The emergent wetland plant *Typha* demonstrated its ability for the phytoremediation of clofibric acid–contaminated waters (Dordio et al., 2009). Another common wetland plant, *Phragmites australis*, showed its phytoremediation efficiency for ciprofloxacin, oxytetracycline, and sulfamethazine (Liu et al., 2013). Similarly, 13 antibiotics were detected in tissues of *Salvinia natans, Hydrocharis dubia,* and *Ceratophyllum demersum*, collected at Baiyangdian Lake, China, and thus may be tested or screened for phytoremediation technology (Li et al., 2012). Wetland plants like *Myriophyllum aquaticum* (parrot feather) and *Pistia stratiotes* (water lettuce) were used to study the phytoremediation of tetracycline and oxytetracycline in aqueous media, and root exudates (consisting of enzymes and metabolites) played a significant role in the degradation of these emerging contaminants of concern (Gujarathi et al., 2005). Further, the toxicity of other PPCPs, like diclofenac, ibuprofen, and acetaminophen in horseradish (*Armoratia rusticana*) and flax (*Linum usitatissimum*), an agro-crop, was investigated in hydroponic culture (Kotyza et al., 2010).

Wetland plant–based reactors may provide a future option for the phytoremediation of emerging contaminants of particular concern, like ibuprofen and fluoxetine. To this end, duckweed reactors (*Landoltia punctata* and *Lemna minor*) offer promising future prospects (Reinhold et al., 2010). Various emerging contaminants under the category of PPCPs, that is, diclofenac, naproxen, ibuprofen, caffeine, and clofibric acid, interacted with important and potent wetland plants, like *Salvinia molesta, L. minor, Ceratophyllum demersum,* and *Elodea canadensis,* under microcosms, and it was observed that caffeine was easily remediated by the free-floating plants *L. minor* and *S. molesta* (>99%), followed by *E. canadensis* (95%) and *C. demersum* (83%), during the 38-day investigation period. On the other hand, another PPCP, ibuprofen, was removed more competently in reactors planted with the submerged *E. canadensis* (78%), a wetland of particular relevance (Matamoros et al., 2012b). It is worth mentioning in this context that the photodegradation or removal of PPCPs may also result in certain intermediates, and their phyto-removal should also be taken into consideration (Matamoros et al., 2012b). Similarly, other research has reported the complete removal (100%) of ibuprofen from the nutrient medium by poplar cells (Iori et al., 2012). Recently, first-hand information on remediation of azole pesticides (tebuconazole and imazalil) with the wetland plant *Phragmites australis* is investigated with 96–99% removal efficiency. Further, in this context, it was observed through enantiomeric study that *R*-imazalil was remediated faster in this wetland plant when compared with *S*-imazalil (Ferreira et al., 2017b). Moreover, *Spartina maritime*, a wetland plant, removed target PPCPs in constructed wetlands, in conjunction with wastewater treatment plants, through the mechanism of adsorption to plant roots and sediments and rhizoremediation (Ferreira et al., 2017a). Also, the phytoremediation of the anthelmintics drug praziquantel in *Phragmites australis* biomass in a recent study on constructed wetlands can be considered an important research finding in effluent treatment in the agriculture, domestic, and industrial sectors (Maršík et al., 2017).

Table 1.2 lists the phytoremediation cases of PPCPs (emerging contaminants of concern) with wetland plants.

PHYTOREMEDIATION OF PATHOGENIC MICROBES AND ITS MECHANISM IN CONSTRUCTED WETLANDS

In view of various waterborne diseases by pathogenic microbes, a group of phytotechnologists diverted or shifted their interest toward their phytoremediation (Martin et al., 2012; Yan et al., 2016; Alufasi et al., 2017). To this end, screened potent wetland plants, like emergent *Typha dominguensis* and *Typha latifolia*, *Phragmites australis*, *Cyperus papyrus*, and *Echinochloa pyramidalis*, were tested for pathogen removal, and interestingly, the extent of phytoremediation was plant species specific, which varied spatially and temporally (Ritcher and Weaver, 2003; Martin et al., 2012; Abdel-Shafy and El-Khateeb, 2013; Giacoman-Vallejos et al., 2015; Yan et al., 2016; Alufasi et al., 2017).

In this context, it is worth mentioning that constructed wetlands are efficient in eliminating diverse microbial pathogens, including bacteria (fecal coliform [FC] bacteria and Enterobacteriaceae), viruses, and protozoan cysts (Greenway, 2005) using wetland plants like *Glyceria* and *Phragmites* (Ottová et al., 1997) or *Typha angustifolia* (Khatiwada and Polprasert, 1999).

Pertaining to mechanisms involved in the remediation of pathogenic microbes, physical (filtration of microorganisms through the rhizosphere, attachment to the substrate, and sedimentation) and chemical (attenuation due to ultraviolet radiation) technologies have certain constraints which paved the way to biological or phytoremediation through wetland plants. The extent of remediation of pathogenic microbes greatly depends on the composition and construction of the hydraulic regime, retention time, and density of wetland plants in constructed wetlands. Rhizospheres of rooted wetland plants of constructed wetlands are rich in exudates of diverse composition. Biotic and abiotic stresses in the rhizospheric environment of wetland plants release exudates that lead to complex and intricate wetland plant root–soil–microbe relations. Antimicrobial activities of the exudates adversely affect or remove pathogenic microbes and retain beneficial microbes that favorably alter the physical and chemical properties of the rhizospheric environment (Bais et al., 2006).

Some plants, like sweet basil (*Ocimum basilicum*), secrete exudates when exposed to specific molecules that stimulate a response (elicitors). Such plants secrete antimicrobial exudates (rosmaric acid) when infected by *Pseudomonas aeruginosa*. Further, *Arabidopsis thaliana*, on infestation with *Pseudomonas syringae*, secretes peroxidases for defense, while this compound is not released in *Arabidopsis thaliana* on infestation with *Sinorhizobium meliloti*. These results imply that the release of exudates in the rhizospheric environment of wetland plants may be microbial pathogen specific (Haichar et al., 2014). Further, it has been observed that proteins like

TABLE 1.2

Phytoremediation Cases of PPCPs from Different Potent Wetland Plants

S. No.	Wetland Plants	PPCPs as Emerging Contaminants	Research Undertaken by Group
1.	*Typha, Salvinia molesta, Elodea canadensis, Ceratophyllum demersum, Lemna minor* (spath, lettuce)	Clofibric acid, diclofenac	Matamoros et al. (2012b)
2.	*Typha, Salvinia molesta, Elodea canadensis, Ceratophyllum demersum, Lemna minor* (spath, lettuce)	Ibuprofen, caffeine, naproxen	Matamoros et al. (2012b), Dordio et al. (2011a), Calderón-Preciado et al. (2011)
3.	*Typha* spp.	Clofibric acid, ibuprofen, carbamazepine	Dordio et al. (2009, 2011a,b)
4.	*Zannichellia palustris*	Sulfathiazole, sulfamethazine, sulfamethoxazole, sulfapyridine, tetracycline, oxytetracycline	Garcia-Rodríguez et al. (2013)
5.	*Pistia stratiotes* (water lettuce)	Tetracycline, oxytetracycline	Gujarathi et al. (2005)
6.	*Salvinia molesta*	Diclofenac, naproxen, ibuprofen, caffeine, clofibric acid	Matamoros et al. (2012a)
7.	*Scirpus validus*	Diclofenac, carbamazepine, naproxen	Zhang et al. (2012, 2013)
8.	*Phragmites autralis*	Diclofenac, ibuprofen, acetaminophen, enrofloxacin, tetracycline, enrofloxacin	Kotyza et al. (2010), Carvalho et al. (2012)
9.	*Myriophyllum aquaticum* (parrot feather)	Tetracycine, oxytetracycline	Gujarathi et al. (2005)
10.	*Lemna minor*	Flumequine, ibuprofen, fluoxetine, clofibric acid	Forni et al. (2001), Cascone et al. (2004), Reinhold et al. (2010)
11.	*Lemna gibba*	Sulfamethoxazole	Brain et al. (2008)
12.	*Landoltia punctate* (duckweed)	Ibuprofen, fluoxetine, clofibric acid	Reinhold et al. (2010)
13.	*Elodea canadensis*	Diclofenac, naproxen, ibuprofen, caffeine, clofibric acid	Matamoros et al. (2012b)
14.	*Ceratophyllum demersum*	Diclofenac, naproxen, ibuprofen, caffeine, clofibric acid	Matamoros et al. (2012b)
15.	*Azolla filicuolides*	Sulfadimethoxine	Forni et al. (2002)

(Continued)

TABLE 1.2 (CONTINUED)
Phytoremediation Cases of PPCPs from Different Potent Wetland Plants

S. No.	Wetland Plants	PPCPs as Emerging Contaminants	Research Undertaken by Group
16.	*Armoratia rusticana* (horseradish)	Diclofenac, ibuprofen, acetaminophen	Kotyza et al. (2010)
17.	*Chrysopogon zizanioides* (vetiver grass)	Tetracycline	Datta et al. (2013)
18.	*Cyperus alternifolius*	PhACs	Yan et al. (2016)
19.	*Phragmites australis*	Anthelmintics drug praziquantel	Maršík et al. (2017)
20.	*Spartina maritime*	PPCPs	Ferreira et al. (2017a)
21.	*Phragmites australis*	Tebuconazole and imazalil	Ferreira et al. (2017b)

lectins, secreted by plants, initiate the formation of tight microbial aggregates on the root surface, enhancing pathogen removal from wastewater.

Several global case studies have demonstrated the role of wetland plants in the phytoremediation of pathogenic microbes. Pilot scale as well as laboratory-scale artificially constructed wetlands built with *Phragmites australis* and *Typha* spp. in Cheshire, United Kingdom, and San Luis Potosi, Mexico, 35–95% and 35–90% for total coliform (TC) and FC, respectively, in planted beds (Rivera et al., 1995). Further, in horizontal flow wetlands in New Zealand comprising *Schoenoplectus validus*, 76.2%–95.2% of faecal coliforms was removed in 7 days from dairy wastewater (Tanner et al., 1995). In northwestern Spain, 78.39%–94.8% of TC, *Escherichia coli*, fecal streptococci, and *Clostridium perfringes* coliphages was removed in a subsurface flow wetland built with *Typha latifolia* in *Salix atrocinerea* when treating sewage effluent (Reinoso et al., 2008). Likewise, the wetland plants *Eichornia crassipes* and *T. latifolia* in a constructed wetland of Ranchi, India, remediated fecal bacteria from effluent mixed with sewage and industrial waste, removing 87%–95% and 24%–71% during the dry and rainy seasons, respectively (Hazra et al., 2011). A horizontal surface flow-constructed wetland with *Echinochloa pyramidalis* in western Cameroon remediated FCs and fecal streptococci in the range of 86%–92% (Martin et al., 2012). Moreover, the pathogenic microbe *E. coli* was removed in the range of 2.8–4 log units in aerated wetlands and 0.8 log units in gravel-based vertical flow beds in subsurface flow-constructed wetlands built with *P. australis* in Germany and India (Headley et al., 2013; Makvana and Sharna, 2013). Similarly, a subsurface flow horizontal-constructed wetland comprising *Typha dominguensis* and *T. latifolia* in the Yucatan Peninsula, Mexico, led to 70%–83% and 65%–78% TC and FC removal, respectively, after 24 h and 80%–82% and 86%–91% TC and FC removal, respectively, after 48 h (Giacoman-Vallejos et al., 2015). In constructed wetlands, to avoid large limitations on land use, biohedge water hyacinth wetlands are frequently used for the remediation of diverse emerging contaminants. Sand filtration may be a very useful mechanism to eliminate emerging contaminants in general and pathogenic microbes in particular (Diaz, 2016).

Phytoremediation of Pharmaceutical Products or Antibiotics with Wetland Plants

The last decade has witnessed an increased number of phytoremediation studies on pharmaceutical products and antibiotics in view of their adverse global environmental and human health impacts as one of the notable emerging contaminants. Unfortunately, as these pharmaceuticals are not completely metabolized in the body, excessive amounts of pharmaceuticals and their metabolites have been introduced into the aquatic environment through sewage, which carries the excreta of individuals who have used these chemicals and agricultural runoff comprising livestock manure (Zhang et al., 2017). One recent research investigation revealed that enrofloxacin and ceftiofur (veterinary antibiotics) did not influence or affect livestock wastewater remediation with constructed wetland-removing metals. Even in the presence of these veterinary antibiotics, the metal removal efficiency of constructed wetland was in the range of 75%–99% (Almeida et al., 2017).

Moreover, it is worth mentioning that these pharmaceuticals and their derivatives are not efficiently eliminated through sewage treatment plants, and thus phytoremediation through wetland plants may be a sustainable option in this context. Constructed wetlands (comprising wetland plants) may be an efficient tool in removing pharmaceuticals (Verlicchi and Zambello, 2014). To this end, a key issue in improving the removal of pharmaceuticals by constructed wetlands is to develop an understanding of their kinetics (Kadlec and Wallace, 2008; Zhang et al., 2017). *Typha latifolia*, *Phragmites australis*, *Iris pseudacorus*, *Juncus effusus*, and *Berula erecta* are potent wetland plants in constructed wetlands in phytoremediation studies on pharmaceutical products or antibiotics (Zhang et al., 2017). In a constructed wetland of China, *Cyperus alternifolius* interaction with pharmaceutically active compounds (PhACs) has been demonstrated through molecular mechanisms (an integrated biochemical and proteomic analysis), and it was noted that reactive oxygen species could be effectively counteracted by enhanced antioxidant enzyme activities, and therefore the photosynthetic pigments were ultimately restored (Yan et al., 2016).

Thus, wetland plants possess immense potential for phytoremediation of emerging contaminants in constructed wetlands. However, the uptake and metabolism of emerging contaminants by plants here is poorly understood due to the lack of good analytics; therefore, in a recent study by Petrie et al. (2017), the first methodology was developed and validated for the multiresidue determination of 81 micropollutants or emerging contaminants (pharmaceuticals, PCPs, and illicit drugs) using microwave-accelerated extraction in the emergent macrophyte *Phragmites australis*.

REFERENCES

Abdel-Shafy HI, El-Khateeb MA. 2013. Integration of septic tank and constructed wetland for the treatment of wastewater in Egypt. *Desalin Water Treat* 51:3539–3546.

Afzal M, Khan QM, Sessitsch A. 2014. Endophytic bacteria: Prospects and applications for the phytoremediation of organic pollutants. *Chemosphere* 117C:232–242.

Al-Baldawi IA et al. 2017. Bioaugmentation for the enhancement of hydrocarbon phytoremediation by rhizobacteria consortium in pilot horizontal subsurface flow constructed wetlands. *Int J Environ Sci Technol* 14:75–84.

Almeida CM et al. 2017. Constructed wetlands for the removal of metals from livestock wastewater—Can the presence of veterinary antibiotics affect removals? *Ecotoxicol Environ Saf* 137:143–148.

Alufasi R, Gere J, Chakauya E, Lebea P, Parawira W, Chingwaru W. 2017. Mechanisms of pathogen removal by macrophytes in constructed wetlands. *Environ Technol Rev* 6(1):135–144.

Amon JP et al. 2007. Development of wetland constructed for the treatment of groundwater contaminated by chlorinated ethenes. *Ecol Eng* 30:51–66.

Anning AK, Korsah PE, Addo-Fordjour P. 2013. Phytoremediation of wastewater *with Limnocharis flava, Thalia geniculata* and *Typha latifolia* in constructed wetlands. *Int J Phytoremediation* 15(5):452–464.

Bacci D, Pietrini F, Bertolotto P, Pérez P, Barcelò D, Zacchini M, Donati E. 2017. Response of *Lemna gibba* L. to high and environmentally relevant concentrations of ibuprofen: Removal, metabolism and morpho-physiological traits for biomonitoring of emerging contaminants. *Sci Total Environ* 584–585:363–373.

Bais HP, Weir TL, Perry LG, Gilroy S, Vivanco JM. 2006. The role of root exudates in rhizosphere interactions with plants and other organisms. *Annu Rev Plant Biol* 57:233–266.

Baldantoni D, Alfani A, Di Tommasi P, Bartoli G, De Santo A. 2004. Assessment of macro and microelement accumulation capability of two aquatic plants. *Environ Pollut* 130:149–156.

Baldantoni D, Maisto G, Bartoli G, Alfani A. 2005. Analyses of three native aquatic plant species to assess spatial gradients of lake trace element contamination. *Aquat Bot* 83:48–60.

Bankston JL et al. 2002. Degradation of trichloroethylene in wetland microcosms containing broad-leaved cattail and eastern cottonwood. *Water Res* 36:1539–1546.

Bartrons M, Peñuelas J. 2017. Pharmaceuticals and personal-care products in plants. *Trends Plant Sci* 22(3):194–203.

Bertrand M, Poirier I. 2005. Photosynthetic organisms and excess of metals. *Phtosynthetica* 43:345–353.

Best EPH et al. 1999. Environmental behavior of explosives in groundwater from the Milan army ammunition plant in aquatic and wetland plant treatments. Removal, mass balances and fate in groundwater of TNT and RDX. *Chemosphere* 38:3383–3396.

Braeckevelt M, Mirschel G, Wiessner A, Rueckert M, Reiche N, Vogt C, Schultz A, Paschke H, Kuschk P, Kaestner M. 2008. Treatment of chlorobenzene-contaminated groundwater in a pilot-scale constructed wetland. *Ecol Eng* 33:45–53.

Braeckevelt M, Reiche N, Trapp S, Wiessner A, Paschke H, Kuschk P, Kaestner M. 2011. Chlorobenzene removal efficiencies and removal processes in a pilot-scale constructed wetland treating contaminated groundwater. *Ecol Eng* 37:903–913.

Brain RA, Ramirez AJ, Fulton BA, Chambliss CK, Brooks BW. 2008. Herbicidal effects of sulfamethoxazole in *Lemna gibba*: Using paminobenzoic acid as a biomarker of effect. *Environ Sci Technol* 42(23):8965–8970.

Calderón-Preciado D, Jiménez-Cartagena C, Matamoros V, Bayona JM. 2011. Screening of 47 organic microcontaminants in agricultural irrigation waters and their soil loading. *Water Res* 45(1):221–231.

Campos VM, Merino I, Casado R, Pacios LF, Gómez L. 2008. Review: Phytoremediation of organic pollutants. *Span J Agric Res* 6:38–47.

Carvalho Y et al. 2017. Inclusion complex between β-cyclodextrin and hecogenin acetate produces superior analgesic effect in animal models for orofacial pain. *Biomed Pharmacother* 93:754–762.

Carvalho PN, Basto M, Almeida C, Brix H. 2014. A review of plant–pharmaceutical interactions: From uptake and effects in crop plants to phytoremediation in constructed wetlands. *Environ Sci Pollut Res* 21:11729–11763.

Carvalho PN, Basto MCP, Almeida CMR. 2012. Potential of *Phragmites australis* for the removal of veterinary pharmaceuticals from aquatic media. *Bioresour Technol* 116(0):497–501.

Carvalho PN, Rodrigues PNR, Evangelista R, Basto MCP, Vasconcelos MTSD. 2011. Can salt marsh plants influence levels and distribution of DDTs in estuarine areas? *Estuar Coast Shelf Sci* 93:415–419.

Cascone A, Forni C, Migliore L. 2004. Flumequine uptake and the aquatic duckweed, *Lemna minor* L. *Water Air Soil Pollut* 156(1):241–249.

Chatterjee S, Chattopadhyay B, Datta S, Mukhopadhyay SK. 2004. Possibility of heavy metal remediation in east Calcutta wetland ecosystem using selected plant species. *J Ind Leather Tech Assoc* 4:299–311.

Chatterjee S, Chetia M, Singh L, Chattopadyay B, Datta S, Mukhopadhyay SK. 2011. A study on the phytoaccumulation of waste elements in wetland plants of a Ramsar site in India. *Environ Monit Assess* 178:361–371.

Chow KL, Man YB, Tam NFY, Liang Y, Wong MH. 2017. Removal of decabromodiphenyl ether (BDE-209) using a combined system involving TiO_2 photocatalysis and wetland plants. *J Hazard Mater* 322:263–269.

Daneshvar A, Svanfelt J, Kronberg L, Weyhenmeyer GA. 2012. Neglected sources of pharmaceuticals in river water—Footprints of a Reggae festival. *J Environ Monitor* 14(2):596–603.

Darajeh N et al. 2016. Modeling BOD and COD removal from palm oil mill secondary effluent in floating wetland by *Chrysopogon zizanioides* (L.) using response surface methodology. *J Environ Manag* 181:343–352.

Datta R, Das P, Smith S, Punamiya P, Ramanathan DM, Reddy R, Sarkar D. 2013. Phytoremediation potential of vetiver grass [*Chrysopogon zizanioides* (L.)] for tetracycline. *Int J Phytoremediation* 15(4):343–351.

Daughton CG, Ternes TA. 1999. Pharmaceuticals and personal care products in the environment: Agents of subtle change? *Environ Health Perspect* 107(6):907–944.

Diaz PM. 2016. Constructed wetlands and water hyacinth macrophyte as a tool for wastewater treatment: A review. *J Adv Civ Eng* 2(1):1–12.

Doni S, Macci C, Peruzzi E. 2015. Heavy metal distribution in a sediment phytoremediation system at pilot scale. *Ecol Eng* 81:146–157.

Dordio A, Ferro R, Teixeira D, Palace AJ, Pinto AP, Dias CMB. 2011a. Study on the use of *Typha* spp. for the phytotreatment of water contaminated with ibuprofen. *Int J Environ Anal Chem* 91(7–8):654–667.

Dordio AV et al. 2009. Toxicity and removal efficiency of pharmaceutical metabolite clofibric acid by *Typha* spp.—Potential use for phytoremediation? *Bioresource Technol* 100(3):1156–1161.

Dordio AV, Carvalho AJP. 2013. Organic xenobiotics removal in constructed wetlands, with emphasis on the importance of the support matrix. *J Hazard Mater* 252–253:272–292.

Doumett S, Fibbi D, Azzarello E, Mancuso S, Mugnai S, Petruzzelli G, Del Bubba M. 2010. Influence of the application renewal of glutamate and tartrate on Cd, Cu, Pb and Zn distribution between contaminated soil and *Paulownia tomentosa* in a pilot-scale assisted phytoremediation study. *Int J Phytoremediation* 13:1–17.

Du L, Liu W. 2012. Occurrence, fate, and ecotoxicity of antibiotics in agro-ecosystems. A review. *Agron Sustain Dev* 32(2):309–327.

Dugan PR. 1972. *Biochemical Ecology of Water Pollution*. New York: Plenum.

Dushenkov V, Kapulnik Y. 2000. Phytofiltration of metals. In *Phytoremediation of Toxic Metals—Using Plants to Clean Up the Environment*, ed. I Raskin, BD Ensley, 89–106. New York: John Wiley & Sons.

Dutta R et al. 2013. Phytoremediation potential of vetiver grass [*Chrysopogon zizanioides* (L.)] for tetracycline. *Int J Phytoremediat* 15(4):343–351.

Elless MP, Poynton CY, Willms CA, Doyle MP, Lopez AC, Sokkary DA, Ferguson BW, Blaylock MJ. 2005. Pilot-scale demonstration of phytofiltration for treatment of arsenic in New Mexico drinking water. *Water Res* 39:3863–3872.

Etim EE. 2012. Phytoremediation and its mechanisms: A review. *Int J Environ Bioenergy* 21:20–36.

Farraji H, Zaman NQ, Tajuddin RM, Faraji H. 2016. Advantages and disadvantages of phytoremediation: A concise review. *Int J Environ Technol Sci* 2:69–75.

Ferreira A et al. 2017a. Remediation of pharmaceutical and personal care products (PPCPs) in constructed wetlands: Applicability and new perspectives. In *Phytoremediation*, ed. A. Ansari et al., 277–292. Cham, Switzerland: Springer.

Ferreira A et al. 2017b. Enantioselective uptake, translocation and degradation of the chiral pesticides tebuconazole and imazalil by *Phragmites australis*. *Environ Pollut* 229:362–370.

Forni C, Cascone A, Cozzolino S, Migliore L. 2001. Drugs uptake and degradation by aquatic plants as a bioremediation technique. *Minerva Biotecnol* 13(2):151–152.

Forni C, Cascone A, Fiori M, Migliore L. 2002. Sulphadimethoxine and *Azolla filiculoides* Lam.: A model for drug remediation. *Water Res* 36(13):3398–3403.

Fountoulakis MS, Terzakis S, Kalogerakis N, Manios T. 2009. Removal of polycyclic aromatic hydrocarbons and linear alkylbenzene sulfonates from domestic wastewater in pilot constructed wetlands and a gravel filter. *Ecol Eng* 35:1702–1709.

Garcia-Rodríguez A, Matamoros V, Fontàs C, Salvadó V. 2013. The influence of light exposure, water quality and vegetation on the removal of sulfonamides and tetracyclines: A laboratory-scale study. *Chemosphere* 90(8):2297–2302.

Gbogbo F, Otoo S. 2015. The concentrations of five heavy metals in components of an economically important urban coastal wetland in Ghana: Public health and phytoremediation implications. *Environ Monit Assess* 187:655.

Gerhardt KE, Huang X-D, Glick BR, Greenberg BM. 2009. Phytoremediation and rhizoremediation of organic soil contaminants: Potential and challenges. *Plant Sci* 176(1):20–30.

Giacoman-Vallejos G, Ponce-Caballelo C, Champaigne P. 2015. Pathogen removal from swine and domestic wastewater by experimental constructed wetlands. *Water Sci Technol* 71:1263–1270.

Gordon M, Choe N, Duffy J, Ekuan G, Heilman P, Muiznieks I, Ruszaj M, Shurtleff BB, Strand S, Wilmoth J, Newman LA. 1998. Phytoremediation of trichloroethylene with hybrid poplars. *Environ Health Perspect* 106:1001–1004.

Gortio AM et al. 2017. A review on the application of constructed wetlands for the removal of priority substances and contaminants of emerging concern listed in recently launched EU legislation. *Environ Pollut* 227:428–443.

Grassi M, Rizzo L, Farina A. 2013. Endocrine disruptors compounds, pharmaceuticals and personal care products in urban wastewater: Implications for agricultural reuse and their removal by adsorption process. *Environ Sci Pollut Res* 20(6):3616–3628.

Greenway M. 2005. The role of constructed wetlands in secondary effluent treatment and water reuse in subtropical and arid Australia. *Ecol Eng* 25:501–509.

Gujarathi NP, Haney BJ, Linden JC. 2005. Phytoremediation potential of *Myriophyllum aquaticum* and *Pistia stratiotes* to modify antibiotic growth promoters, tetracycline, and oxytetracycline, in aqueous wastewater systems. *Int J Phytoremediation* 7(2):99–112.

Haichar FZ et al. 2014. Root exudates mediated interactions belowground. *Soil Biol Biochem* 77:69–80.

Haritash AK, Kaushik CP. 2009. Biodegradation aspects of polycyclic aromatic hydrocarbons (PAHs): A review. *J Hazard Mater* 169:1–15.

Hazra M, Avishek K, Pathak G. 2011. Developing an artificial wetland system for wastewater treatment: A designing perspective. *Int J Environ Prot* 1:8–18.

Headley T et al. 2013. *Escherichia coli* removal and internal dynamics in subsurface flow ecotechnologies: Effects of design and plants. *Ecol Eng* 61:564–574.

Herath I, Vithanage M. 2015. Phytoremediation in constructed wetlands. In *Phytoremediation: Management of Environmental Contaminants*, ed. AA Ansari et al., 243. Vol. 2. Cham, Switzerland: Springer.

Hyland KC et al. 2015. Accumulation of contaminants of emerging concern in food crops: Part I. Edible strawberries and lettuce grown in reclaimed water. *Environ Toxicol Chem* 34:2213–2221.

Imfeld G, Braeckevelt M, Kuschk P, Richnow HH. 2009. Monitoring and assessing processes of organic chemicals removal in constructed wetlands. *Chemosphere* 74:349–362.

Iori V, Pietrini F, Zacchini M. 2012. Assessment of ibuprofen tolerance and removal capability in *Populus nigra* L. by in vitro culture. *J Hazard Mater* 229–230(0):217–223.

Jiang J-Q, Zhou Z, Sharma VK. 2013. Occurrence, transportation, monitoring and treatment of emerging micro-pollutants in waste water—A review from global views. *Microchem J* 110:292–300.

Kadlec RH, Cuvellier C, Stober T. 2010. Performance of the Columbia, Missouri, treatment wetland. *Ecol Eng* 36:672–684.

Kadlec RH, Wallace S, 2008. *Treatment Wetlands*. 2nd ed. Boca Raton, FL: CRC Press.

Khatiwada NR, Polprasert C. 1999. Kinetics of fecal coliform removal in constructed wetlands. *Water Sci Technol* 40:109–116.

Kim S, Aga DS. 2007. Potential ecological and human health impacts of antibiotics and antibiotic-resistant bacteria from wastewater treatment plants. *J Toxicol Environ Health B* 10(8):559–573.

Knight R, Kadlec R, Ohlendorf H. 1999. The use of treatment wetlands for petroleum industry effluents. *Environ Sci Technol* 33:973–980.

Kotyza J, Soudek P, Kafka Z, Vaněk T. 2010. Phytoremediation of pharmaceuticals—Preliminary study. *Int J Phytoremediation* 12(3):306–316.

Kumar, PBAN, Dushenkov, V, Motto, H, Raskin, I. 1995. Phytoextraction: The use of plants to remove heavy metals from soils. *Environ Sci Technol* 29(5):1232–1238.

Kumar R, Lee J, Cho J. 2012. Fate, occurrence, and toxicity of veterinary antibiotics in environment. *J Korean Soc Appl Biol Chem* 55(6):701–709.

Kuschk P, Wiessner A, Kappelmeyer U, Weissbrodt E, Kästner M, Stottmeister U. 2003. Annual cycle of nitrogen removal by a pilot-scale subsurface horizontal flow in a constructed wetland under moderate climate. *Water Res* 37:4236–4242.

Lasat MM. 2000. Phytoextraction of metals from contaminated soil: A review plant/soil/metal interaction and assessment of pertinent agronomic issues. *J Hazard Subst Res* 2(5):1–25.

Lee J, Rai PK, Jeon YJ, Kim KH, Kwon EE. 2017. The role of algae and cyanobacteria in the production and release of odorants in water. *Environ Pollut* 227:252–262.

Leroy MC et al. 2015. Assessment of PAH dissipation processes in large-scale outdoor mesocosms simulating vegetated road-side swales. *Sci Total Environ* 520:146–153.

Li W, Shi Y, Gao L, Liu J, Cai Y. 2012. Occurrence of antibiotics in water, sediments, aquatic plants, and animals from Baiyangdian Lake in North China. *Chemosphere* 89(11):1307–1315.

Lin Q, Mendelssohn I. 2009. Potential of restoration and phytoremediation with *Juncus roemerianus* for diesel-contaminated coastal wetlands. *Ecol Eng* 35:85–91.

Lin Q, Mendelssohn, IA. 2012. Impacts and Recovery of the Deepwater Horizon Oil Spill on Vegetation Structure and Function of Coastal Salt Marshes in the Northern Gulf of Mexico. *Environ Sci Technol* 46(7):3737–3743.

Liu L, Liu Y-H, Liu C-X, Wang Z, Dong J, Zhu G-F, Huang X. 2013. Potential effect and accumulation of veterinary antibiotics in *Phragmites australis* under hydroponic conditions. *Ecol Eng* 53(0):138–143.

Lv T, Zhang Y, Zhang L, Carvalho PN, Arias CA, Brix H. 2016a. Removal of the pesticides imazalil and tebuconazole in saturated constructed wetland mesocosms. *Water Res* 91:126–136.

Lv T et al. 2016b. Phytoremediation of imazalil and tebuconazole by four emergent wetland plant species in hydroponic medium. *Chemosphere* 148:459–466.

Lv T et al. 2017. Functionality of microbial communities in constructed wetlands used for pesticide remediation: Influence of system design and sampling strategy. *Water Res* 110:241–251.

Makvana KS, Sharna MK. 2013. Assessment of pathogen removal potential of root zone technology from domestic wastewater. *Univ J Environ Res Technol* 3:401–406.

Manousaki E, Kalogerakis N. 2011. Halophytes—An emerging trend in phytoremediation. *Int J Phytoremediation* 13:959–969.

Maršík P et al. 2017. Study of praziquantel phytoremediation and transformation and its removal in constructed wetland. *J Hazard Mater* 323(5):394–399.

Martin L et al. 2012. Removal of fecal bacteria and nutrients from domestic wastewater in a horizontal surface flow wetland vegetated with *Echinochloa pyramidalis*. *Afr J Environ Sci Technol* 6:337–345.

Matamoros V, Arias CA, Nguyen LX, Salvadó V, Brix H. 2012a. Occurrence and behavior of emerging contaminants in surface water and a restored wetland. *Chemosphere* 88(9):1083–1089.

Matamoros V, Bayona JM. 2006. Elimination of pharmaceuticals and personal care products in subsurface flow constructed wetlands. *Environ Sci Technol* 40(18):5811–5816.

Matamoros V, Nguyen LX, Arias CA, Salvadó V, Brix H. 2012b. Evaluation of aquatic plants for removing polar microcontaminants: A microcosm experiment. *Chemosphere* 88(10):1257–1264.

McIntyre T. 2003. Phytoremediation of heavy metals from soils. *Adv Biochem Eng Biotechnol* 78:97–123.

Medina VF, Mccutcheon SC. 1996. Phytoremediation: Modeling removal of TNT and its breakdown products. *Remediation J* 7:31–45.

Michałowicz J. 2014. Bisphenol A—Sources, toxicity and biotransformation. *Environ Toxicol Pharmacol* 37:738–758.

Migliore L, Cozzolino S, Fiori M. 2000. Phytotoxicity to and uptake of flumequine used in intensive aquaculture on the aquatic weed, *Lythrum salicaria* L. *Chemosphere* 40(7):741–750.

Moore MT, Schulz R, Cooper CM, Smith JRS, Rodgers JH Jr. 2002. Mitigation of chlorpyrifos runoff using constructed wetlands. *Chemosphere* 46:827–835.

Opuene K, Okafor EC, Agbozu E. 2008. Partitioning characteristics of heavy metals in a nontidal freshwater ecosystem. *Int J Environ Res* 2(3):285–290.

Ottová V, Balcarová J, Vymazal J. 1997. Microbial characteristics of constructed wetlands. *Water Sci Technol* 35:117–123.

Pandey SK, Tripathi BD, Prajapati SK, Mishra VK, Upadhyay AR, Rai PK, Sharma AP. 2005. Magnetic properties of vehicle derived particulates and amelioration by *Ficus infectoria*: A keystone species. *Ambio* 34(8):645–646.

Petrie B, Smith BD, Youdan J, Barden R, Kasprzyk-Hordern B. 2017. Multi-residue determination of micropollutants in *Phragmites australis* from constructed wetlands using microwave assisted extraction and ultra-high-performance liquid chromatography tandem mass spectrometry. *Anal Chim Acta* 959:91–101.

Phillips LA, Greer CW, Farrell RE, Germida JJ. 2012. Plant root exudates impact the hydrocarbon degradation potential of a weathered-hydrocarbon contaminated soil. *Appl Soil Ecol* 52:56–64.

Pomati F, Netting AG, Calamari D, Neilan BA. 2004. Effects of erythromycin, tetracycline and ibuprofen on the growth of *Synechocystis* sp. and *Lemna minor*. *Aquat Toxicol* 67(4):387–396.

Rai PK. 2007a. Phytoremediation of Pb and Ni from industrial effluents using *Lemna minor*: An eco-sustainable approach. *Bull Biosci* 5(1):67–73.

Rai PK. 2007b. Wastewater management through biomass of *Azolla pinnata*: An ecosustainable approach. *Ambio* 36(5):426–428.

Rai PK. 2008a. Phytoremediation of Hg and Cd from industrial effluents using an aquatic free floating macrophyte *Azolla pinnata*. *Int J Phytoremediation* 10(5):430–439.

Rai PK. 2008b. Mercury pollution from chlor-alkali industry in a tropical lake and its biomagnification in aquatic biota: Link between chemical pollution, biomarkers and human health concern. *Human Ecol Risk Assess Int J* 14:1318–1329.

Rai PK. 2009a. Heavy metal phytoremediation from aquatic ecosystems with special reference to macrophytes. *Crit Rev Environ Sci Technol* 39(9):697–753.

Rai PK. 2009b. Comparative assessment of soil properties after bamboo flowering and death in a tropical forest of Indo-Burma hot spot. *Ambio* 38(2):118–120.

Rai PK. 2010a. Microcosm investigation on phytoremediation of Cr using *Azolla pinnata*. *Int J Phytoremediation* 12:96–104.

Rai PK. 2010b. Phytoremediation of heavy metals in a tropical impoundment of industrial region. *Environ Monit Assess* 165:529–537.

Rai PK. 2010c. Seasonal monitoring of heavy metals and physico-chemical characteristics in a lentic ecosystem of sub-tropical industrial region, India. *Environ Monit Assess* 165:407–433.

Rai PK. 2010d. Heavy metal pollution in lentic ecosystem of sub-tropical industrial region and its phytoremediation. *Int J Phytoremediation* 12(3):226–242.

Rai PK. 2011. *Heavy Metal Pollution and Its Phytoremediation through Wetland Plants*. New York: Nova Science Publisher.

Rai PK. 2012a. Assessment of multifaceted environmental issues and model development of an Indo-Burma hot spot region. *Environ Monit Assess* 184:113–131.

Rai PK. 2012b. An eco-sustainable green approach for heavy metals management: Two case studies of developing industrial region. *Environ Monit Assess* 184:421–448.

Rai PK. 2013a. Environmental magnetic studies of particulates with special reference to biomagnetic monitoring using roadside plant leaves. *Atmos Environ* 72:113–129.

Rai PK. 2013b. *Plant Invasion Ecology: Impacts and Sustainable Management*. New York: Nova Science Publisher.

Rai PK. 2015a. Paradigm of plant invasion: Multifaceted review on sustainable management. *Environ Monit Assess* 187:759–785.

Rai PK. 2015b. What makes the plant invasion possible? Paradigm of invasion mechanisms, theories and attributes. *Environmental Skeptics Critics* 4(2):36–66.

Rai PK. 2015c. Concept of plant invasion ecology as prime factor for biodiversity crisis: Introductory review. *Int Res J Environ Sci* 4(5):85–90.

Rai PK. 2015d. *Environmental Issues and Sustainable Development of North East India*. Saarbrücken, Germany: Lambert Academic Publisher.

Rai PK. 2016a. Biodiversity of roadside plants and their response to air pollution in an Indo-Burma hotspot region: Implications for urban ecosystem restoration. *J Asia Pac Biodivers* 9:47–55.

Rai PK. 2016b. Impacts of particulate matter pollution on plants: Implications for environmental biomonitoring. *Ecotoxicol Environ Saf* 129:120–136.

Rai PK, Chutia BM. 2016a. Particulate matter bio-monitoring through magnetic properties of an Indo-Burma hotspot region. *Chem Ecol* 32(6):550–574.

Rai PK, Chutia B. 2016b. Biomagnetic monitoring through *Lantana* leaves in an Indo-Burma hot spot region. *Environ Skeptics Critics* 5(1):1–11.

Rai PK, Chutia BM, Patil SK. 2014. Monitoring of spatial variations of particulate matter (PM) pollution through bio-magnetic aspects of roadside plant leaves in an Indo-Burma hot spot region. *Urban For Urban Green* 13:761–770.

Rai PK, Mishra A, Tripathi BD. 2010. Heavy metals and microbial pollution of river Ganga: A case study on water quality at Varanasi. *Aquat Ecosyst Health Manag* 13(4):352–361.

Rai PK, Panda LS. 2014. Dust capturing potential and air pollution tolerance index (APTI) of some roadside tree vegetation in Aizawl, Mizoram, India: An Indo-Burma hot spot region. *Air Quality Atmos Health* 7(1):93–101.

Rai PK, Panda LS, Chutia BM, Singh MM. 2013. Comparative assessment of air pollution tolerance index (APTI) in the industrial (Rourkela) and non industrial area (Aizawl) of India: An eco-management approach. *Afr J Environ Sci Technol* 7(10):944–948.

Rai PK, Singh M. 2016. *Eichhornia crassipes* as a potential phytoremediation agent and an important bioresource for Asia Pacific region. *Environ Skeptics Critics* 5(1):12–19.

Rai PK, Singh MM. 2015. *Lantana camara* invasion in urban forests of an Indo-Burma hotspot region and its ecosustainable management implication through biomonitoring of particulate matter. *J Asia Pac Biodivers* 8:375–381.

Rai PK, Tripathi BD. 2008. Heavy metals in industrial wastewater, soil and vegetables in Lohta village, India. *Toxicol Environ Chem* 90(2):247–257.

Rai PK et al. 2018. A critical review of ferrate(VI)-based remediation of soil and groundwater. *J Environ Res* 160:420–448.

Rane NR, Chandanshive VV, Watharkar AD, Khandare RV, Patil TS, Pawar PK, Govindwar SP. 2015. Phytoremediation of sulfonated Remazol Red dye and textile effluents by *Alternanthera philoxeroides*: An anatomical, enzymatic and pilot scale study. *Water Res* 83:271–281.

Ranieri E, Gikas P, Tchobanoglous G. 2013. BTEX removal in pilot-scale horizontal subsurface flow constructed wetlands. *Desalination Water Treat* 51:3032–3039.

Reinhold D, Vishwanathan S, Park JJ, Oh D, Saunders FM. 2010. Assessment of plant-driven removal of emerging organic pollutants by duckweed. *Chemosphere* 80(7):687–692.

Reinoso R, Torres LA, Becares E. 2008. Efficiency of natural systems for removal of bacteria and pathogenic parasites from wastewater. *Sci Total Environ* 395:80–86.

Reyes-Contreras C, Matamoros V, Ruiz I, Soto M, Bayona JM. 2011. Evaluation of PPCPs removal in a combined anaerobic digester-constructed wetland pilot plant treating urban wastewater. *Chemosphere* 84:1200–1207.

Rezg R, El-Fazaa S, Gharbi N, Mornagui B. 2014. Bisphenol A and human chronic diseases: Current evidences, possible mechanisms, and future perspectives. *Environ Int* 64:83–90.

Ribeiro H, Mucha AP, Almeida CMR, Bordalo AA. 2014. Potential of phytoremediation for the removal of petroleum hydrocarbons in contaminated salt marsh sediments. *J Environ Manage* 137:10–15.

Ritcher AY, Weaver RW. 2003. Treatment of domestic wastewater by subsurface flow constructed wetlands filled with gravel and tire chip media. *Environ Technol* 24:1561–1567.

Rivera F et al. 1997. The application of the root zone method for the treatment and reuse of high strength abattoir waste in Mexico. *Water Sci Technol* 35:271–278.

Roy S, Hanninen O. 1994. Pentachlorophenol: Uptake/elimination kinetics and metabolism in an aquatic plant, *Eichhornia crassipes*. *Environ Toxicol Chem* 13:763–773.

Samecka-Cymerman A, Stepien D, Kempers AJ. 2004. Efficiency in removing pollutants by constructed wetland purification systems in Poland. *J Toxicol Environ Health A* 67(4):265–275.

Sauvêtre, A, Schröder, P. 2015. Uptake of carbamazepine by rhizomes and endophytic bacteria of *Phragmites australis*. *Front Plant Sci* 6:1–11.

Seeger EM, Kuschk P. 2011. Bioremediation of benzene-, MTBE- and ammonia-contaminated groundwater with pilot-scale constructed wetlands. *Environ Pollut* 159:3769–3776.

Singh OV, Labana S, Pandey G, Budhiraja R, Jain RK. 2003. Phytoremediation: An overview of metallic ion decontamination from soil. *Appl Microbiol Biotechnol* 61(5–6):405–412.

Singh MM, Rai PK. 2016. Microcosm investigation of Fe (iron) removal using macrophytes of Ramsar lake: A phytoremediation approach. *Int J Phytoremediation* 18(12):1231–1236.

Sood A, Uniyal PP, Prasanna R, Ahluwalia AS. 2012. Phytoremediation potential of aquatic macrophyte, *Azolla. Ambio* 41:122–137.

St-Cyr L, Campbell PGC. 1994. Bioavailability of sediment-bound metals for *Vallisneria Americana* Michx, a submerged aquatic plant, in the St. Lawrence River. *Can J Fish Aquat Sci* 57:1330–1341.

Stearman GK. 2003. Pesticide removal from container nursery runoff in constructed wetland cells. *J Environ Qual* 32:1548–1556.

Stottmeister U et al. 2003. Effects of plants and microorganisms in constructed wetlands for wastewater treatment. *Biotechnol Adv* 22:93–117.

Sui Q et al. 2015. Occurrence, sources and fate of pharmaceuticals and personal care products in the groundwater: A review. *Emerg Contam* 1:14–24.

Sun G, Zhao Y, Allen S, Cooper D. 2006. Generating "tide" in pilot-scale constructed wetland to enhance agricultural wastewater treatment. *Eng Life Sci* 6(6):560–565.

Susarla S, Medina VF, McCutcheon SC. 2002. Phytoremediation: An ecological solution to organic chemical contamination. *Ecol Eng* 18:647–658.

Syranidou E, Christofilopoulos S, Kalogerakis N. 2017a. Juncus spp.—The helophyte for all (phyto)remediation purposes? *New Biotechnol.* doi: 10.1016/j.nbt.2016.12.005.

Syranidou E, Christofilopoulos S, Politi M, Weyens N, Venieri D, Vangronsveld J, Kalogerakis N. 2017b. Bisphenol-A removal by the halophyte *Juncus acutus* in a phytoremediation pilot: Characterization and potential role of the endophytic community. *J Hazard Mater* 323:350–358. http://dx.doi.org/doi:10.1016/j.jhazmat.2016.05.034.

Tanner CC, Clayton JS, Upsdell MP. 1995. Effect of loading rate and planting on treatment of dairy farm wastewaters in constructed wetlands. I. Removal of oxygen demand, suspended solids and faecal coliforms. *Water Res* 29:17–26.

Ternes TA. 1998. Occurrence of drugs in German sewage treatment plants and rivers. *Water Res* 32(11):3245–3260.

Terzakis S et al. 2008. Constructed wetlands treating highway runoff in the central Mediterranean region. *Chemosphere* 72:141–149.

Tiwari S, Dixit S, Verma N. 2007. An effective means of biofiltration of heavy metal contaminated water bodies using aquatic weed *Echhornia crassipes. Environ Monit Assess* 129:253–256.

Verlicchi P et al. 2013. Removal of selected pharmaceuticals from domestic wastewater in an activated sludge system followed by a horizontal subsurface flow bed—Analysis of their respective contributions. *Sci Total Environ* 454–455:411–425.

Verlicchi P, Zambello E. 2014. How efficient are constructed wetlands in removing pharmaceuticals from untreated and treated urban wastewaters? A review. *Sci Total Environ* 470:1281–1306.

Vymazal J, Brezinova T. 2015. The use of constructed wetlands for removal of pesticides from agricultural runoff and drainage: A review. *Environ Int* 75:11–20.

Wallace S, Kadlec R. 2005. BTEX degradation in a cold-climate wetland system. *Water Sci Technol* 51(9):165–172.

Wang Q, Kelly BC. 2017. Occurrence, distribution and bioaccumulation behaviour of hydrophobic organic contaminants in a large-scale constructed wetland in Singapore. *Chemosphere* 183:257–265.

Wetzel RG. 1983. *Limnology.* 2nd ed. Philadelphia: Saunders College Publishing.

Williams JB. 2002. Phytoremediation in wetland ecosystems: Progress, problems, and potential. *Crit Rev Plant Sci* 21:607–635.

Yan Q et al. 2016. Insights into the molecular mechanism of the responses for *Cyperus alternifolius* to PhACs stress in constructed wetlands. *Chemosphere* 164:278–289.

Yeh T, Chou C, Pan C. 2009. Heavy metal removal within pilot-scale constructed wetlands receiving river water contaminated by confined swine operations. *Desalination* 249:368–373.

Zarate FM, Schulwitz SE, Stevens KJ, Venables BJ. 2012. Bioconcentration of triclosan, methyl-triclosan, and triclocarban in the plants and sediments of a constructed wetland. *Chemosphere* 88:323–329.

Zhang DQ et al. 2013. Carbamazepine and naproxen: Fate in wetland mesocosms planted with *Scirpus validus*. *Chemosphere* 91(1):14–21.

Zhang DQ, Hua T, Gersberg RM, Zhu J, Ng WJ, Tan SK. 2012. Fate of diclofenac in wetland mesocosms planted with *Scirpus validus*. *Ecol Eng* 49:59–64.

Zhang Y, Lv T, Carvalho PN, Zhang L, Arias CA, Chen Z, Brix H. 2017. Ibuprofen and iohexol removal in saturated constructed wetland mesocosms. *Ecol Eng* 98:394–402.

2 Phytoremediation: Concept, Principles, Mechanisms, and Applications

INTRODUCTION

In the 1990s, there was considerable interest in developing sustainable, cost-effective technologies for the remediation of emerging contaminant–polluted soil and water (Lasat, 2000). The removal of emerging contaminants using living organisms has recently been attracting a lot of public attention and research and development (R&D) spending (Rai, 2009). The use of plants for the remediation of emerging contaminants offers an attractive alternative, because it is solar driven and can be carried out *in situ*, minimizing cost and human exposure (Salt et al., 1998). Eccles (1999), through a cost–benefit analysis, proved that biological processes for emerging contaminant removal, specifically of heavy metals, are cheaper than conventional technologies.

GREEN SUSTAINABLE TECHNOLOGY—PHYTOREMEDIATION: CONCEPT AND PRINCIPLES

The term *phytoremediation*, which consists of *phyto* in Greek (meaning "plant") and *remedium* in Latin (meaning "correct evil") (Erakhrumen and Agbontalor, 2007), was coined in 1991 (Sachdeva and Sharma, 2012). Phytoremediation is defined as the use of plants and their associated microbes to extract, sequester, and/or detoxify various kinds of environmental pollution or emerging contaminants from water, sediments, soils, and air (as magnetic particles lying in particulate matter) (Memon and Schröder, 2009; Rai et al., 2013, 2014, 2018; Rai, 2013, 2016; Rai and Panda, 2014; Rai and Singh, 2015; Rai and Chutia, 2016). It is the plant-based green technology that received increasing attention after the discovery of hyperaccumulating plants, which are able to accumulate, translocate, and concentrate high amounts of hazardous elements in the harvestable parts (Rahman and Hasegawa, 2011). Phytoremediation is a relatively new approach to the cost-effective treatment of wastewater, groundwater, and soils contaminated with organic xenobiotics, heavy metals, and radionuclides (Gardea-Torresdey, 2003; Gardea-Torresdey et al., 2004, 2005). While there are numerous descriptions of the term, phytoremediation can be summed up with one clear definition: the use of plants for the removal of pollutants from the environment

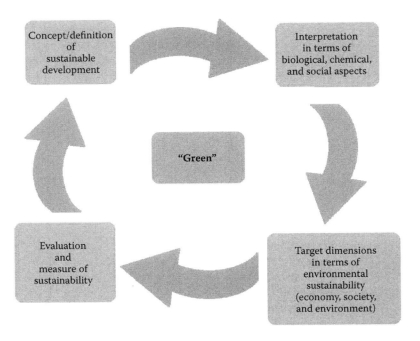

FIGURE 2.1   Sustainable development with phytotechnologies: changing paradigm.

(Gardea-Torresdey, 2003). It is worth mentioning that phytoremediation is a green technology having the potential to attain a sustainability paradigm (Figure 2.1).

The effectiveness of phytoremediation requires well-planned strategies for the decontamination process. Not all plants can develop in contaminated environments. The first step for the use of phytoremediation is to identify species that, besides being suitable to local conditions, are tolerant to contaminants. The second step is to evaluate the capacity of the plant to promote decontamination (Marques et al., 2011; Preussler et al., 2014). The plant used for the phytoremediation should have some characteristic features (Sharma et al., 2014):

1. Fast growing
2. High metal tolerance
3. Resistant to diseases, pests, and so forth
4. Dense root and shoot system (Couselo et al., 2012)
5. Unattractive to animals so that there should be minimum transfer of metals to higher trophic levels of the terrestrial food chain (Bruce et al., 2003)
6. Easy to cultivate and harvest

Phytoremediation tends to detoxify the soil and groundwater compartment of the environment by wetland plants through

1. Alteration of the physical and chemical properties of polluted soils
2. Releasing root exudates, thereby increasing organic carbon

3. Ameliorating aeration by releasing oxygen to the root zone, as well as increasing the porosity of the upper soil zones
4. Intercepting and retarding the movement of contaminants or chemicals
5. Effecting co-metabolic microbial and plant enzymatic transformations of xenobiotics and recalcitrant chemicals
6. Decreasing vertical and lateral migration of contaminants to groundwater by extracting available water and reversing the hydraulic gradient (Susarla et al., 2002; Rai, 2009; Singh and Rai, 2016)

MICROBIAL ASSOCIATION AND PHYTOREMEDIATION OF EMERGING CONTAMINANTS

Most plants are always associated with microorganisms that cause them diseases. They are called endophytes. Endophytes include endophytic bacteria, endophytic fungi, and endophytic ectinomycetes (Raghukumar, 2008; Shehzadi et al., 2016). Not all the endophytes cause diseases. Some of them help plants indirectly. Their activity inside the plants helps and increases their metabolism, forming a beneficial association between them. Different mechanisms have been developed by bacteria to avoid toxicity, such as the following (Pavel et al., 2013; Ullah et al., 2014; Shehzadi et al., 2016; Al-Baldawi et al., 2017):

1. Active efflux pumps
2. Intra-extracellular sequestration
3. Exclusion through permeable barriers
4. Reduction through enzymes
5. Reduction of cellular sensitivity

Endophytes help plants to enhance growth through phytohormone production; supply nitrogen after the nitrogen fixation process; resist environmental stresses (heat, cold, drought, and salt); produce important medicinal, agricultural, and industrial compounds; and enhance phytoremediation after improving the uptake of contaminants and degradation of several toxins (Khan and Doty, 2011; Shehzadi et al., 2016). They are resistant to emerging contaminants and capable of degrading the contaminants.

The heavy metal–resistant endophytes belong to a wide range of taxa; in bacteria, these include *Arthrobacter, Bacillus, Clostridium, Curtobacterium, Enterobacter, Leifsonia, Microbacterium, Paenibacillus, Pseudomonas, Xanthomonadaceae, Staphylococcus, Stenotrophomona,* and *Sanguibacte,* and in fungi, *Microphaeropsis, Mucor, Phoma, Alternaria, Peyronellaea, Steganosporium,* and *Aspergillus* (Li et al., 2012).

The endophytes are densely colonized inside the plant roots, decreasing from the stem to the leaves (Porteous-Moore et al., 2006). In an important study, 41 culturable endophytic bacteria (*Bacillus* [39%], *Microbacterium* [12%], and *Halomonas* [12%]) were isolated from the roots and shoots of three wetland plants, *Typha domingensis, Pistia stratiotes,* and *Eichhornia crassipes,* and identified through 16S rRNA gene sequencing in the context of textile effluents consisting of emerging contaminants and their phytoremediation. In comparison with other rhizosphere microorganisms, endophytes interact more closely with their host plants and can more efficiently improve phytoremediation (Zhang et al., 2011). Among the endophytic genera, Burkholderiaceae, Enterobacteriaceae, and Pseudomonadaceae are the most common cultivable species found (Khan and Doty, 2011). Efficient phytoremediation of total petroleum hydrocarbon (TPH) effluent in HF-type constructed wetlands with *Scirpus grossus* assisted with microbial consortium (Al-Baldawi et al., 2017). Al-Baldawi et al. used a consortium of three rhizobacteria strains (*Bacillus aquimaris*, *Bacillus anthracis*, and *Bacillus cereus*) to augment the wetland plant–based phytoremediation.

PHYTOREMEDIATION MECHANISMS IN WETLAND PLANTS FOR DIVERSE EMERGING CONTAMINANTS

The phytoremediation mechanism for emerging contaminants consists of several processes, such as phytoextraction, rhizofiltration, phytostabilization, phytovolatilization, and phytotransformation or phytodegradation.

Each of the processes haa different role in the accumulation and remediation of the emerging contaminants.

1. Phytoextraction: Phytoextraction is the use of plants to remove emerging contaminants from aquatic bodies by accumulating them in their tissue, particularly in the harvestable roots and the shoots (Kumar et al., 1995), and different procedures of 'phytomining' may lead to recovery of metals from biomass (Erakhrumen and Agbontalor, 2007). Some hyperaccumulator plants absorb unusually large amounts of emerging contaminants, like metals, compared with other plants and the ambient metal concentration (Padmavathiamma and Li, 2007). The process of phytoextraction involves the accumulation of emerging contaminants, like heavy metal liquefaction, transforming them into relatively stable metal fractions (oxidizable and residual fractions), leading to decreased bioavailability and eco-toxicity of heavy metals (Yuan et al., 2011). Two basic important strategies of phytoextraction have developed after several approaches (Salt et al., 1995, 1997). Phytoextraction comprises
    a. Chelate-assisted phytoextraction, or induced phytoextration, in which artificial chelates are added to increase the mobility and uptake of metal contaminants.
    b. Continuous phytoextraction, in which the removal of metal depends on the natural ability of the plant to remediate. Only the number of plant growth repetitions is controlled.
2. Rhizofiltration: Rhizofiltration is the absorption of emerging contaminants by macrophyte roots from the aquatic bodies, precipitate, and concentrate in their biomass (Dushenkov et al., 1995; Erdei et al., 2005). Wetland plants perform the rhizofiltration process to accumulate and concentrate emerging contaminants or metals. The process involves raising plants hydroponically and transplanting them into metal-polluted waters, where plants absorb and concentrate the emerging contaminants,

specifically metals, in their roots and shoots (Dushenkov et al., 1995; Salt et al., 1995; Flathman and Lanza, 1998). The process involves chemisorption, complexation, ion exchange, microprecipitate, hydroxide condensation onto the biosurface, and surface adsorption (Gardea-Torresdey et al., 2004). The efficiency of these processes can be increased by using plants with a heightened ability to absorb and translocate metals (Zhu et al., 1999). Wetlands plants such as *Eichhornia crassipes, Lemna minor,* and *Azolla pinnata* are being used for rhizofiltration for the treatment of contaminated wetland (Rai, 2009). Rhizofiltration is a particularly cost-competitive technology in the treatment of surface or ground water containing relatively low concentrations of toxic metals using wetland plants (Salt et al., 1995).

3. Phytostabilization: Phytostabilization is the immobilization of emerging contaminants through the use of tolerant wetland plants (Salt et al., 1995; Rai, 2009) by absorption and accumulation in their tissues, adsorption in their roots, or precipitation within the root space, stopping the transfer of the contaminants in soil, along with their movement by erosion and deflation (Erdei et al., 2005). It is not intended to remove emerging contaminants from a site, but rather to stabilize them by accumulation in root zones, reducing the mobility of contaminants and preventing migration to groundwater or air, and also reducing the risk to human health and the environment (Padmavathiamma and Li, 2007). For the process of phytostabilization, the appropriate selection of macrophyte is required. Characteristics of wetland plants include tolerance to high levels of contaminants of concern; high production of root biomass able to immobilize these contaminants through uptake, precipitation, or reduction; and retention of applicable contaminants in roots, as opposed to transfer to shoots, to avoid special handling and disposal of shoots (Padmavathiamma and Li, 2007).

4. Photovolatilization: Phytovolatilization is the transpiration of absorbed contaminants by the plants in modified form to the atmosphere (Erdei et al., 2005) from their foliage (Lone et al., 2008). Plants can volatilize certain metals, such as highly toxic mercury (Hg^{2+}) and methylmercury, by reducing them, using the enzyme mercuric reductase, to less harmful Hg (Yadav et al., 2010). The mechanisms/steps of phytovolatilization basically operate in the aerial parts of the macrophyte, such as the leaves (Mukhopadhyay and Maiti, 2010).

5. Phytodegradation: Phytodegradation is the elimination or degradation of metals by a macrophyte's enzyme or enzyme cofactors (Susarla et al., 2002) and its associated microbes (Garbisu and Alkorta, 2001; Garbisu et al., 2002). Enzymes such as dehalogenases, oxygenases, and reductases degrade the compounds that contain emerging contaminants inside the plant bodies (Black, 1995). This is also known as phytotransformation (Mani and Kumar, 2014). Degradation may also occur outside the plant, due to the release of compounds that cause the transformation (Mukhopadhyay and Maiti, 2010). With help of the associated microbes in the root zone

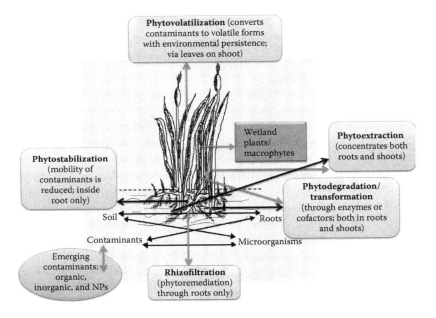

FIGURE 2.2 Generalized mechanics involved in heavy metal hyperaccumulation and phytoremediation of emerging contaminants by wetland plants.

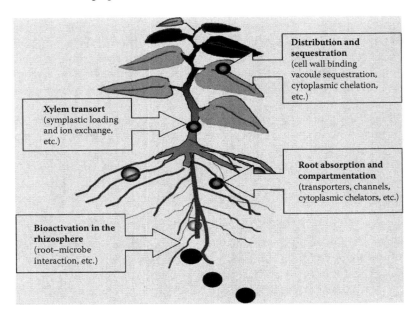

FIGURE 2.3 Relevance of different physiological and biochemical mechanisms in the phytoremediation of emerging contaminants. (From Yang et al., 2005; Rai, P.K., *Crit. Rev. Environ. Sci. Technol.*, 39(9), 697–753, 2009; Rai, P.K., *Heavy Metal Pollution and Its Phytoremediation through Wetland Plants*, Nova Science Publisher, New York, 2011.)

(rhizosphere) of the plant, rhizodegradation also occurs. For a graphical overview of the aforesaid mechanisms and physiological alterations, see Figures 2.2 and 2.3.

ROLE OF ENZYMES IN PHYTOREMEDIATION OF EMERGING CONTAMINANTS

Enzymes in wetland plants possess the potential to detoxify emerging contaminants, specifically organics, as mentioned below.

1. Phosphatase enzyme present in wetland plants, for example, giant duckweed (*Spirodela polyrhiza*) used to break the phosphate chemical bond in organophosphate pesticides
2. Nitroreductase enzyme present in wetland plants; for example, hybrid poplar (*Populus* spp.), stonewort (*Nitella* spp.), and parrot feather (*Myriophyllum aquaticum*) tend to reduce nitro groups on explosives and other nitroaromatic compounds and remove nitrogen from ring structures
3. Nitrilase present in wetland plants like willow (*Salix* spp.), which is used to break cyanide groups from aromatic rings
4. Laccase present in wetland plants like stonewort and parrot feather (*Myriophyllum aquaticum*), which is used to break aromatic rings after TNT is reduced to triaminotoluene
5. Dehalogenase present in wetland plants like hybrid poplar (*Populus* spp.) and several algae spp., such as parrot feather (*Myriophyllum aquaticum*), which tend to dehalogenate chlorinated solvents
6. Peroxidase enzyme present in a unique wetland plant, horseradish (*Armoracia rusticana*)

Molecular or biochemical mechanisms related to the remediation of emerging contaminants, like metals in algae (phycoremediation) and wetland plants (phytoremediation), are demonstrated in Figure 2.4.

UTILITY OF WETLAND PLANTS IN PHYTOREMEDIATION OF EMERGING CONTAMINANTS

Plants are the most tolerant to pollution, which makes them very useful for new emerging environmental biotechnology—phytoremediation (Gawronski et al., 2011). Many wetland plants species are successfully used for the phytoremediation of emerging contaminants in contaminated water bodies (Darajeh et al., 2016). These wetland aquatic plants and their associated microbes were utilized to absorbe and degrade the metals to prevent further contamination of the water bodies. The adequate restoration of these environments requires cooperation, integration, and assimilation of such biotechnological advances, along with traditional and ethical wisdom to unravel the mystery of nature in the emerging field of bioremediation and phytoremediation (Mani and Kumar, 2014). Also, advances in molecular studies with wetland plants are the need of the hour, as they can enhance the efficiency of phytoremediation (Ali et al., 2013).

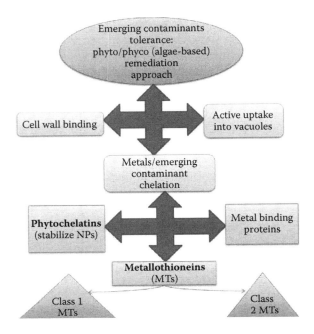

FIGURE 2.4 Molecular or biochemical mechanisms related to the remediation of emerging contaminants: algae-based remediation (phycoremediation) and wetland plants (phytoremediation).

The emerging contaminant and metal uptake and distribution within the plants is affected by some factors (Susarla et al., 2002):

- Physical and chemical properties of the compound (e.g., water solubility, vapor pressure, molecular weight, and octanol–water partition coefficient, K_{OW})
- Environmental characteristics (e.g., temperature, pH, organic matter, and soil moisture content)
- Plant characteristics (e.g., type of root system and type of enzymes)

Wetland plants are extremely beneficial to wetlands because they provide food and shelter for fish and aquatic invertebrates (Rai, 2008b, 2009, 2011, 2012). They also produce oxygen, which helps in overall lake functioning, and provide food for some fish and other wildlife. Macrophytes are considered important components of the aquatic ecosystem, not only because they are a food source for aquatic invertebrates, but also because they act as an efficient accumulator of heavy metals (Rai, 2008b, 2011, 2012). They are unchangeable biological filters and play an important role in the maintenance of the aquatic ecosystem. Aquatic wetland plants are taxonomically closely related to terrestrial plants, but they are aquatic phanerogams, which live in a completely different environment. Their characteristics in accumulating emerging contaminants and metals make them interesting research subjects for testing and modeling ecological theories on evolution and plant succession, as well as on nutrient and metal cycling (Fostner and Whittman, 1979; Rai, 2011, 2012). Therefore, it is very important to understand the functions of wetland plants in aquatic ecosystems. The use of aquatic vascular plants for emerging contaminant phytoremediation is very much emphasized these days for the treatment of industrial effluents before discharge into aquatic ecosystems. Since only aquatic plants can flourish in aquatic environments, naturally requiring simple mineral nutrients and sunlight, they can be conveniently tested for their phytoremediation potential. Aquatic wetland plants, for example, water hyacinth (*Eichhornia crassipes*), water velvet (*Azolla pinnata*), and duckweeds (*Lemna minor* and *Spirodela polyrhiza*), are prevalent in lakes, rivers, and streams all over the globe. A huge amount of money is invested to remove them from polluted aquatic bodies to conserve their aesthetic character and suitably maintain them for the eco-tourism industry (Rai, 2009, 2011, 2012). Wastewater treatment wetlands harboring rich growth of macrophytes can be both a cost-efficient and effective means to improve water quality before effluents are discharged into major rivers (Hammer, 1992, 1994; Kadlec and Knight, 1996; Verhoeven and Mueleman, 1999; Mitsch and Gosselink, 2000; Nzengya and Wishitemi, 2001; Shutes, 2001; Stone et al., 2004; Nahlik and Mitsch, 2006; Rai, 2011, 2012; Singh and Rai, 2016).

Other economic benefits, such as vegetation for animal feed (e.g., floating aquatic plants) and habitat for harvestable fish (e.g., *Tilapia*), make wetlands an attractive option for meeting water quality standards through nutrient reduction for farmers and small industry (Greenway and Simpson, 1996; Denny, 1997; Rai, 2011, 2012). Free-floating macrophytes provide shading of the water column, thereby providing a cooler habitat for fish and macroinvertebrates in what otherwise would be a warm water tropical environment. Despite floating aquatic plants providing cooler water temperatures and abundant food sources that help create an optimal habitat structure for fish and invertebrates, it has been reported that wetlands with floating aquatics support larger mosquito larvae populations than do open-water areas due to reduced dissolved oxygen concentrations (Greenway et al., 2003; Rai, 2011, 2012; Singh and Rai, 2016).

Water hyacinth (*Eichhornia crassipes*) is one of the most commonly used plants in constructed wetlands because of its fast growth rate and large uptake of nutrients and contaminants (Rai, 2007a, 2011, 2012; Rai and Singh, 2016; Singh and Rai, 2016). It has been studied for its tendency to bioaccumulate and biomagnify the heavy metal contaminants present in water bodies (Tiwari et al., 2007; Rai and Singh, 2016). It accumulates metals, and as the recycling process is run by photosynthetic activity and biomass growth, it is sustainable process and cost-efficient (Garbisu et al., 2002; Lu et al., 2004; Bertrand and Poirier, 2005). It has the capacity to accumulate metals such as Cd, Cu, Pb, and Zn in its root tissues (Nor, 1990). However, it has been reported that the growth of water hyacinth poses a problem in the functioning of constructed wetlands due to its exotic invasive nature and rapid decomposition in comparison with other plants (Khan et al., 2000; Rai, 2011, 2012; Yang et al., 2014; Singh and Rai, 2016). Nevertheless, my work on invasion and exotic invasive plants contradicts this limiting issue, in which such invasive waste biomass may act as a potent phytotechnology tool for treating emerging contaminants lying in various environmental compartments (Rai, 2012, 2013, 2015a–d, 2016; Rai and Chutia, 2016; Rai and Singh, 2016).

Macrophytes and wetland plants, including *Eichhornia crassipes*, *Pistia stratiotes*, and *Spirodela polyrrhiza*, remove Cr from wastewater (Mishra and Tripathi, 2008). Duckweeds (family Lemnaceae) appear to be the better alternative and have been recommended for industrial effluent treatment, as they are (1) more tolerant to cold conditions than water hyacinth, (2) more easily harvested than algae, and (3) capable of rapid growth (Sharma and Gaur, 1995; Rai, 2011, 2012; Singh and Rai, 2016). The biomass of aquatic wetland plants, for example, *Azolla*, may be used for recycling municipal wastewater for irrigation by filtering the heavy metals and other pollutants, which assists in water resource conservation (Rai, 2007b, 2011, 2012; Singh and Rai, 2016). At the same time, the waste biomass of wetland plants produced after the treatment can be used for biogas production in an effort toward achieving sustainability in the energy sector. *Azolla pinnata*, endemic to India, after treatment with HCl or HNO_3 to remove heavy metals absorbed, may be of tremendous biofertilizer value due to its association with a cyanobacterium, *Anabena azollae* (Rai, 2007b, 2011, 2012; Singh and Rai, 2016).

Typha species also act as nutrient pumps, absorbing a large amount of nutrients from the sediment and accumulating them in the aboveground tissue (Sharma, 2007). *Typha augustifolia* shows higher metal levels in roots than aboveground tissues (Wu et al., 2014). In constructed wetlands, *Typha domingensis* may be a potent wetland plant in the phytoremediation of extremely deleterious mercury (Gomes et al., 2014). *T. domingensis* is use in constructed wetlands for the enhancement of water quality in water treatment systems (El-Sheikh et al., 2010; Hegazy et al., 2011). It can assess a significant concentration of Cu, Cd, Mg, and ash during its growing season (Eid et al, 2012). Cattail (*Typha latifolia*) and common reed (*Phragmites australis*) have been used successfully for the phytoremediation of Pb and Zn mine tailings under waterlogged conditions (Ye et al., 2004). Polyculture constructed wetlands with *T. latifolia* and *P. australis* may be used to treat mine effluent, particularly boron (Turker et al., 2013). Estuarine sediments colonized by *P. australis* and *Juncus maritimus* were spiked with Cd in the absence and presence of an autochthonous microbial consortium resistant to the metal (Nunes da Silva et al., 2014). They did not have significant signs of toxicity in the increased Cd uptake (Nunes da Silva et al., 2014). Certain metals, like Zn, Fe, Pb, Cu, Ni, Cd, and Cr, were reported to accumulate in higher concentrations inside the root, followed by the leaf and stems, of *Phalaris arundinacea* (reed canarygrass), thus making it a potent bioresource for trace metal removal from bottom sediments (Polechonska and Klink, 2014). Both roots and shoots of *Eleocharis acicularis* effectively phytoremediated indium, Ag, Pb, Cu, Cd, and Zn (Ha et al., 2011).

RHIZOFILTRATION PROCESS INVOLVED IN ACCUMULATION THROUGH WETLAND PLANTS

Wetland plants mostly follow the mechanism of rhizofiltration for the phytoremediation of emerging contaminants. Rhizofiltration is the phytoremediative technique designed for the removal of emerging contaminants and metals from an aquatic environment using wetland plants (Rai, 2009, 2011, 2012; Singh and Rai, 2016). Therefore, roots are thought to be important for element uptake in free-floating macrophytes

(Sharma and Gaur, 1995; Rai, 2011, 2012; Singh and Rai, 2016). Although aquatic wetland plants have been found to accumulate emerging contaminants and metals in their shoots (Greger, 1999; Fritioff et al., 2005; Singh and Rai, 2016), but whether these metals originate from direct uptake from the water or from root-to-shoot trans-location is in most cases an open question.

Root exudates and changes in rhizosphere pH may also cause metals to precipi-tate on root surfaces. As they become saturated with the metal contaminants or other emerging contaminants, roots or whole plants are harvested for disposal (Zhu et al., 1999; Rai, 2011, 2012; Singh and Rai, 2016). Rhizofiltration offers a cost advantage in water treatment because of the ability of plants to remove up to 60% of their dry weight as toxic metals, thus markedly reducing the generation and disposal cost of the hazardous residue. Hence, rhizofiltration is a particularly cost-competitive tech-nology in the treatment of surface or ground water containing relatively low con-centrations of toxic metals (Salt et al., 1995; Rai, 2011, 2012; Singh and Rai, 2016).

Most research works have shown that plants for phytoremediation should accu-mulate emerging contaminants only in the roots (Dushenkov et al., 1995; Salt et al., 1995; Rai, 2011, 2012; Singh and Rai, 2016). Dushenkov and Kapulnik (2000) explained that the translocation of emerging contaminants and metals to the shoot would decrease the efficiency of rhizofiltration by increasing the amount of contami-nated plant residue needing disposal. In contrast, Zhu et al. (1999) suggested that the efficiency of the process can be increased by using plants that have a heightened ability to absorb and translocate metals within the plant. Dushenkov and Kapulnik (2000) described the characteristics of the ideal plants for rhizofiltration. Plants should be able to accumulate and tolerate significant amounts of target metals in conjunction with easy handling, low maintenance cost, and a minimum of second-ary waste disposal. It is also desirable that plants produce significant amounts of root biomass or root surface area (Rai, 2011, 2012; Singh and Rai, 2016).

PHYTOREMEDIATION OF EMERGING CONTAMINANTS WITH WETLAND PLANTS AND MACROPHYTES: EXAMPLES

Various phytotechnologists and water researchers have performed phytoremediation using many wetland plants for the removal of emerging contaminants from aquatic bodies or wetlands. A list of wetland plants used in the phytoremediation of emerg-ing contaminants is given in Table 2.1. Most workers prefer *Eichhornia crassipes* as a common plant for the removal of Pb, Cu, Zn, Hg, Cd, Cr, and Mn (Tiwari et al., 2007; Kumar et al., 2008; Rai, 2009; Rai et al., 2010; Chatterjee et al., 2011; Fawzy et al., 2012; Padmapriya and Murugesan, 2012; Mishra et al., 2013; Sasidharan et al., 2013; Singh and Rai, 2016). It is a well-known fact that *E. crassipes* is invasive and its utilization as a phytoremediating agent and in bioenergy, as well as in nanotech-nology sectors, may aid in its sustainable management.

Wetlands plants of more than 60 species were used for phytoremediation stud-ies by various workers (Rai, 2007a–c, 2009, 2010a–d, 2011, 2012; Rai et al., 2010, 2017; Singh and Rai, 2016). Some of the common macrophytes include *Ipomoea aquatic* (Kumar et al., 2008), *Typha* sp. (Kumar et al., 2008), *Echinochloa* sp. (Kumar et al., 2008), *Hydrilla verticillata* (Kumar et al., 2008; Rai, 2009; Begam

TABLE 2.1
List of Wetland Plants Used for the Phytoremediation of Emerging Contaminants with Their Common Names

Common Name	Scientific Name	Source
Vetiver grass	*Vetiveria zizanioides*	Suelee et al., 2017
Reed	*Phragmites australis*	Rai, 2009; Bonanno, 2013; Li et al., 2013; Ranieri et al., 2013; Turker et al., 2013; Wu et al., 2014; Eid and Shaltout, 2014; Gill et al., 2014; Hechmi et al., 2014; Nunes da Silva et al., 2014; Kumari and Tripathi, 2015; Philippe et al., 2015
Pistia stratiotes (water lettuce), *Eichhornia crassipes* (water hyacinth), *Ipomoea aquatica* (swamp morning glory), *Paspalum repens* (water paspalum), *Azolla microphylla* (Mexican mosquito fern), *Salvinia minima* (water spangles), *Lemna minor* L. (lesser duckweed)	Mentioned	Nahlik and Mitsch, 2006
Tall reed	*Phragmites karka*	Rai, 2009; Turker et al., 2013
Water fern	*Azolla caroliniana*	Rai, 2007b, 2008a–c; Singh and Rai, 2016
Water velvet	*Azolla pinnata*	Rai, 2007b, 2008c, 2010a–d; Rai et al., 2010; Singh and Rai, 2016
Water bloom/algal bloom	*Microcystis* sp.	Rai and Tripathi, 2007a–c; Rai, 2011, 2012; Lee et al. 2017
Cattail	*Typha latifolia*	Rai, 2008b, 2009; Leto et al., 2013; Turker et al., 2013; Kumari and Tripathi, 2015
Lesser Indian reed mace	*Typha angustata*	Chandra and Yadav, 2010
Narrow leaf cattail	*Typha angustifolia*	Yadav et al., 2012; Li et al., 2013; Gill et al., 2014; Gomes et al., 2014
Bulrush	*Typha domingensis*	Hegazy et al., 2011; Eid et al., 2012; Bonanno, 2013; Philippe et al., 2015
Poplar trees	*Populus deltoids*	Rai, 2008b; Yadav et al., 2010
Pondweed/curly leaf	*Potamogeton natans*	Rai, 2008b, 2011
Pondweed	*Potamogeton crispus*	Rai, 2008b; Upadhyay et al., 2014

(Continued)

TABLE 2.1 (CONTINUED)
List of Wetland Plants Used for the Phytoremediation of Emerging Contaminants with Their Common Names

Common Name	Scientific Name	Source
Parrot's feather	*Myriophyllum spicatum*	Rai, 2009
Umbrella plant	*Cyperus alternifolius*	Rai, 2009; Soda et al., 2012; Yadav et al., 2012; Leto et al., 2013; Yang and Ye, 2014
Duckweed	*Lemna minor*	Rai, 2007a
Water hyacinth	*Eichhornia crassipes*	Rai, 2008b, 2009; Chunkao et al., 2012; Malar et al., 2014, 2015; Mazumdar and Das, 2015
Smart weed	*Polygonum hydropiper*	Rai, 2008b
Smooth cordgrass	*Spartina alterniflora*	Rai, 2008b
Water zinnia	*Wedelia trilobata*	Rai, 2008b
Water lettuce	*Pistia stratiotes*	Rai, 2008b; Das et al., 2014
Irish leaved rush	*Juncus xihoides*	Rai, 2009
Hard rush/blue rush	*Juncus inflexus*	Philippe et al., 2015
Fuzzy water clover	*Marsilea dromondii*	Rai, 2009
Reed canarygrass	*Phalaris arundinacea*	Rai, 2009; Polechonska and Klink, 2014
Salt marsh bulrush	*Scirpus robustus*	Rai, 2009
Rabbitfoot grass	*Polypogon monspeliensis*	Rai, 2008a
Needle spikerush/dwarf hairgrass	*Eleocharis acicularis*	Ha et al., 2011
Zebra rush	*Scirpus tabernaemontani*	Rai, 2009
Lanceleaf water plantain	*Alisma lanceolatom*	Philippe et al., 2015
Carice volpina	*Carex cuprina*	Philippe et al., 2015
Great willowherb	*Epilobium hirsutum*	Philippe et al., 2015
Yellow iris flag	*Irish pseudacorus*	Philippe et al., 2015
Soft rush	*Juncus effuses*	Ladislas et al., 2013
Greater pond sedge	*Carex riparia*	Ladislas et al., 2013
Sessile joyweed	*Alternathera sessilis*	Mazumdar and Das, 2015
Prickly amaranth	*Amaranthus spinosus*	Mazumdar and Das, 2015
Red amaranth	*Amaranthus cruentus*	Mazumdar and Das, 2015
Centella	*Centella asiatica*	Mazumdar and Das, 2015
Coco yam	*Colocasia esculenta*	Mazumdar and Das, 2015
Bitter vine	*Mikania micrantha*	Mazumdar and Das, 2015
Little ironweed	*Cyanthillium cinereum*	Mazumdar and Das, 2015
Corn sow thistle	*Sonchus arvensis*	Mazumdar and Das, 2015
Feverfew	*Parthenium hysterophorus*	Mazumdar and Das, 2015
False daisy	*Eclipta prostrate*	Mazumdar and Das, 2015
Lamb's quarters	*Chenopodium album*	Mazumdar and Das, 2015
Water spinach	*Ipomoea aquatica*	Mazumdar and Das, 2015

(Continued)

TABLE 2.1 (CONTINUED)
List of Wetland Plants Used for the Phytoremediation of Emerging Contaminants with Their Common Names

Common Name	Scientific Name	Source
Two rowed rush	*Fimbristylis dichotoma*	Mazumdar and Das, 2015
Bubani	*Fimbristylis bisumbellata*	Mazumdar and Das, 2015
Castor plant	*Ricinus communis*	Mazumdar and Das, 2015
Leucas	*Leucas lavandulifolia*	Mazumdar and Das, 2015
Kidney bean	*Phaseolus vulgaris*	Mazumdar and Das, 2015
Linear leaf water primrose	*Ludwigia hyssopifolia*	Mazumdar and Das, 2015
Bermuda grass	*Cynodon dactylon*	Mazumdar and Das, 2015
Thalia love grass	*Eragrsostis atrovirens*	Mazumdar and Das, 2015
Indian goose grass	*Eleusine indica*	Mazumdar and Das, 2015
American nightshade	*Solanum americanum*	Mazumdar and Das, 2015
Tobacco	*Nicotiana plumbaginifolia*	Mazumdar and Das, 2015
Vegetable fern	*Diplazium esculentum*	Mazumdar and Das, 2015
Acorus	*Acorus tatarinowii*	Yang and Ye, 2014
Alligator weed	*Alternanthera philoxeroides*	Yang and Ye, 2014
Water fern	*Ceratopteris thalictroides*	Yang and Ye, 2014
Water pennywort	*Hydrocotyle vulgaris*	Yang and Ye, 2014
Torpedo grass	*Panicum repens*	Yang and Ye, 2014
Dwarf rotala	*Rotala rotundifolia*	Yang and Ye, 2014
Thyme-leaf speedwell	*Veronica serpyllifolia*	Yang and Ye, 2014
Arum lily	*Zantedeschia aethiopica*	Yang and Ye, 2014
Chinese taro	*Alocasia cucullata*	Yang and Ye, 2014
Amazon sword plant	*Echinodorus amazonicus*	Yang and Ye, 2014
Melon sword plant	*Echinodorus osiris*	Yang and Ye, 2014
Spikerush	*Eleocharis geniculate*	Yang and Ye, 2014
Flat spikerush	*Fimbristylis monostachya*	Yang and Ye, 2014
Water primrose	*Jussiaea linifolia*	Yang and Ye, 2014
Indian paspalum	*Paspalum scrobiculatum*	Yang and Ye, 2014
Frogsmouth	*Philydrum lanuginosum*	Yang and Ye, 2014
Triangular club-rush	*Scirpus triqueter*	Yang and Ye, 2014
Sea rush	*Juncus maritimus*	Nunes da Silva et al., 2014
Giant reed	*Arundo donax*	Bonanno, 2013
Indian shot	*Canna indica*	Yadav et al., 2012
Dark stonewort	*Nitella opaca*	Sooksawat et al., 2013
Parrot's feather	*Myriophyllum aquaticum*	Souza et al., 2013
Soft rush	*Juncus effuses*	Rahman et al., 2014
Sweetflag	*Acorus gramineus*	Soda et al., 2012
Hydrilla	*Hydrilla verticillata*	Xue et al., 2010
Common salvinia	*Salvinia minima*	Ponce et al., 2015
Fennel pondweed	*Potamogeton pectinatus*	Upadhyay et al., 2014
Emergent cattail/other	*Limnocharis flava, Thalia geniculata*, and *Typha latifolia*	Anning et al., 2013

and HariKrishna, 2010), *Nelumbo nucifera* (Kumar et al., 2008), *Vallisneria spiralis* (Kumar et al., 2008; Rai, 2009; Rai and Tripathi, 2009), *Aponogeton natans* (Rai, 2008b), *Cyperus rotundus* (Rai, 2008b), *Ipomoea aquatica* (Rai, 2008b), *Marsilea quadrifolia* (Rai, 2008b), *Potamogeton pecitnatus* (Rai, 2008b), *Lemna* sp. (Rai et al., 2010; Singh and Rai, 2016), *Spirodela polyrhiza, Azolla pinnata* (Rai et al., 2010; Singh and Rai, 2016), *Polygonum* sp. (Bako and Daudu, 2007; Rai, 2009), *Ludwigia* sp. (Bako and Daudu, 2007), *Elodea canadensis* (Begam and HariKrishna, 2010), *Salvinia* sp. (Begam and HariKrishna, 2010; Singh and Rai, 2016), *Wolffia arrhiza* (Chatterjee et al., 2011), *Pistia stratiotes* (Chatterjee et al., 2011; Singh and Rai, 2016), *Trapa bispinosa* (Chatterjee et al., 2011), *Cynodon dactylon* (Chatterjee et al., 2011), *Scirpus* sp. (Chatterjee et al., 2011), *Colocasia esculenta* (Chatterjee et al., 2011), *Sagittaria montevidensis* (Chatterjee et al., 2011), *Schoenoplectus californicus* (Boudet et al., 2011), *Ricciocarpus natans* (Boudet et al., 2011), *Bidens tripartitus* (Branković et al., 2011), *Lycopus europaeus* (Branković et al., 2011), *Myriophyllum* sp. (Fawzy et al., 2012), *Phragmites* sp. (Fawzy et al., 2012), *Spirodela polyrrhiza* (Loveson et al., 2013; Singh and Rai, 2016), *Juncus effuses* (Ladislas et al., 2013), *Carex riparia* (Ladislas et al., 2013), and *Jussiaea repens* (Mishra et al., 2013).

REFERENCES

Al-Baldawi IA et al. 2017. Bioaugmentation for the enhancement of hydrocarbon phytoremediation by rhizobacteria consortium in pilot horizontal subsurface flow constructed wetlands. *Int J Environ Sci Technol* 14:75–84.

Ali H, Khan E, Sajad MA. 2013. Phytoremediation of heavy metals: Concepts and applications. *Chemosphere* 91:869–881.

Anning AK, Korsah PE, Addo-Fordjour P. 2013. Phytoremediation of wastewater with *Limnocharis flava, Thalia geniculata* and *Typha latifolia* in constructed wetlands. *Int J Phytoremediation* 15(5):452–464.

Bako SP, Daudu P. 2007. Trace metal contents of the emergent macrophytes *Polygonum* sp. and *Ludwigia* sp. in relation to the sediments of two freshwater lake ecosystems in the Nigerian savanna. *J Fish Aquat Sci* 2(1):63–70.

Begam A, HariKrishna S. 2010. Bioaccumulation of trace metals by aquatic plants. *Int J Chemtech Res* 2(1):250–254.

Bertrand M, Poirier I. 2005. Photosynthetic organisms and excess of metals. *Phtosynthetica* 43:345–353.

Black H. 1995. Absorbing possibilities: Phytoremediation. *Environ Health Perspect* 103(12):1106–1108.

Bonanno G. 2013. Comparative performance of trace element bioaccumulation and biomonitoring in the plant species *Typha domingensis, Phragmites australis* and *Arundo donax. Ecotoxicol Environ Saf* 97:124–130.

Boudet LC, Escalante A, von Haeften G, Moreno V, Gerpe M. 2011. Assessment of heavy metals accumulation in two aquatic macrophytes: A field study. *J Braz Soc Ecotoxicol* 6(1):57–64.

Branković S, Pavlović-Muratspahić D, Topuzović M, Glišić R, Banković D, Stanković M. 2011. Environmental study of some metals on several aquatic macrophytes. *Afr J Biotechnol* 10(56):11956–11965.

Bruce SL, Noller BN, Grigg AH, Mullen BF, Mulligan DR, Ritchie PJ, Currey N, Ng JC. 2003. A field study conducted at Kidston gold mine, to evaluate the impact of arsenic and zinc from mine tailing to grazing cattle. *Toxicol Lett* 137:23–34.

Chandra R, Yadav S. 2010. Potential of *Typha angustfolia* for phytoremediation of heavy metals from aqueous solution of phenol and melanoidin. *Ecol Eng* 36:1277–1284.
Chatterjee S, Chetia M, Singh L, Chattopadyay B, Datta S, Mukhopadhyay SK. 2011. A study on the phytoaccumulation of waste elements in wetland plants of a Ramsar site in India. *Environ Monit Assess* 178:361–371.
Couselo JL, Corredoira E, Vieitez AM, Ballester A. 2012. Plant tissue culture of fast growing trees for phytoremediation research. *Methods Mol Biol* 877:247–263.
Darajeh N et al. 2016. Modeling BOD and COD removal from palm oil mill secondary effluent in floating wetland by *Chrysopogon zizanioides* (L.) using response surface methodology. *J Environ Manag* 181:343–352.
Das S, Goswami S, Talukdar AD. 2014. A study on cadmium phytoremediation potential of water lettuce, *Pistia stratiotes* L. *Bull Environ Contam Toxicol* 92:169–174.
Denny P. 1997. Implementation of constructed wetlands in developing countries. *Water Sci Technol* 35:27–34.
Dushenkov V, Kapulnik Y. 2000. Phytofiltration of metals. In *Phytoremediation of Toxic Metals: Using Plants to Clean Up the Environment*, ed. I Raskin, BD Ensley, 89–106. New York: John Wiley & Sons.
Dushenkov V, Kumar PBAN, Motto H, Raskin I. 1995. Rhizofiltration—The use of plants to remove heavy metals from aqueous streams. *Environ Sci Technol* 29(5):1239–1245.
Eccles H. 1999. Treatment of metal-contaminated wastes: Why select a biological process? *Trends Biotechnol* 17:462–465.
Eid EM, Shalhout KH, El-Sheikh M, Asaeda T. 2012. Seasonal courses of nutrients and heavy metals in water, sediment and above- and below-ground *Typha domingensis* biomass in Lake Burullus (Egypt): Perspectives for phytoremediation. *Flora* 207:783–794.
Eid EM, Shaltout KH. 2014. Monthly variation of trace elements accumulation and distribution in above- and below-ground biomass of *Phragmites australis* (Cav.) Trin. Ex Steudel in Lake Burullus (Egypt): A biomonitoring application. *Ecol Eng* 73:17–25.
El-Sheikh MA, Saleh HI, Ei-Quosy DE, Mahmoud AA. 2010. Improving water quality in polluted drains with free water surface constructed wetlands. *Ecol Eng* 36:1478–1484.
Erakhrumen A, Agbontalor A. 2007. Review phytoremediation: An environmentally sound technology for pollution prevention, control and remediation in developing countries. *Educ Res Rev* 2(7):151–156.
Erdei L, Mezôsi G, Mécs I, Vass I, Fôglein F, Bulik L. 2005. Phytoremediation as a program for decontamination of heavy-metals polluted environment. *Acta Biol Szeged* 49(1–2):75–76.
Fawzy MA, Badr NES, Khatib AE, Kassem AAE. 2012. Heavy metal biomonitoring and phytoremediation potentialities of aquatic macrophytes in River Nile. *Environ Monit Assess* 184:1753–1771.
Flathman PE, Lanza GR. 1998. Phytoremediation: Current views on an emerging green technology. *J Soil Contam* 7(4):415–435.
Fritioff A, Kautsky L, Greger M. 2005. Influence of temperature and salinity on heavy metal uptake by submersed plants. *Environ. Poll* 133:265–274.
Fostner U, Whittman GTW. 1979. *Metal Pollution in the Aquatic Environment*. Berlin: Springer-Verlag.
Garbisu C, Alkorta I. 2001. Phytoextraction: A cost effective plant-based technology for the removal of metals from the environment. *Bioresour Technol* 77(3):229–236.
Garbisu C, Hernandez-Allica J, Barrutia O, Alkortaand I, Becerril JM. 2002. Phytoremediation: A technology using green plants to remove contaminants from polluted areas. *Rev Environ Health* 17(3):173–188.
Gardea-Torresdey JL. 2003. Phytoremediation: Where does it stand and where will it go? *Environ Progress* 22(1):A2–A3.

Gardea-Torresdey JL, de la Rosa G, Peralta-Videa JR. 2004. Use of phytofiltration technologies in the removal of heavy metals: A review. *Pure Appl Chem* 76(4):801–813.

Gardea-Torresdey JL, Peralta-Videa JR, Rosa GDL, Parson JG. 2005. Phytoremediation of heavy metals and study of the metal coordination by x-ray spectroscopy. *Coord Chem Rev* 17–18:1797–1810.

Gawronski SW, Greger M, Gawronska H. 2011. Plant taxonomy and phytoremediation. In *Detoxification of Heavy Metals (Soil Biology)*, ed. I Sherameti, A Varma, 91–109. Vol. 30. Berlin: Springer.

Gill LW, Ring P, Higgins MNP, Johnston PM. 2014. Accumulation of heavy metals in a constructed wetland treating road runoff. *Ecol Eng* 70:133–139.

Gomes MVT et al. 2014. Phytoremediation of water contaminated with mercury using *Typha domingensis* in constructed wetland. *Chemosphere* 103:228–233.

Greger M. 1999. Metal availability and bioconcentration in plants. In *Heavy Metal Stress in Plants: From Molecule to Ecosystems*, eds. MNV Prasad, J Hagemeyer. Berlin: Springer-Verlag.

Greenway M, Dale P, Chapman H. 2003. An assessment of mosquito breeding and control in four surface flow wetlands in tropical-subtropical Australia. *Water Sci Technol* 48:121–128.

Greenway M, Simpson JS. 1996. Artificial wetlands for wastewater treatment, water reuse and wildlife in Queensland, Australia. *Water Sci Technol* 33:221–229.

Ha NT, Sakakibara M, Sano S. 2011. Accumulation of indium and other heavy metals by *Eleocharis acicularis*: An operation for phytoremediation and phytomining. *Bioresour Technol* 102:2228–2234.

Hammer DA. 1992. Designing constructed wetland systems to treat agricultural non-point source pollution. *Ecol Eng* 1:49–82.

Hammer DA. 1994. *Constructed Wetlands for Wastewater Treatment—Municipal, Industrial and Agriculture*. Chelsea, MI: Lewis Publishers.

Hechmi N, Aissa NB, Abdenaceur H, Jedidi N. 2014. Evaluating the phytoremediation potential of *Phragmites australis* grown in pentachlorophenol and cadmium co-contaminated soils. *Environ Sci Pollut Res* 21:1304–1313.

Hegazy AK, Abdel-Ghani NT, El-Chaghaby GA. 2011. Phytoremediation of industrial wastewater potentiality by *Typha domingensis*. *Int J Environ Sci Technol* 8:639–648.

Kadlec RH, Knight RL. 1996. *Treatment Wetlands*. Boca Raton, FL: Lewis Publishers.

Khan AG, Kuek C, Chaudry TM, Khoo CS, Hayes WJ. 2000. Role of plants, mycorrhizae and phytochelators in heavy metals contaminated land remediation. *Chemosphere* 41:197–207.

Khan Z, Doty S. 2011. Endophyte-assisted phytoremediation. *Curr Opin Plant Biol* 12:97–105.

Kumar JIN, Soni R, Kumar RN, Bhatt I. 2008. Macrophytes in phytoremediation of heavy metal contaminated water and sediments in Periyej Community Reserve, Gujarat, India. *Turk J Fish Aquat Sci* 8:193–200.

Kumar PBAN, Dushenkov V, Motto H, Raskin I. 1995. Phytoextraction—The use of plants to remove heavy metals from soils. *Environ Sci Technol* 29:1232–1238.

Kumari M, Tripathi BD. 2015. Efficiency of *Phragmites australis* and *Typha latifolia* for heavy metal removal from wastewater. *Ecotoxicol Environ Saf* 112:80–86.

Ladislas S, Gérente C, Chazarenc F, Brisson J, Andrès Y. 2013. Performances of two macrophytes species in floating treatment wetlands for cadmium, nickel, and zinc removal from urban stormwater runoff. *Water Air Soil Pollut* 224:1408.

Lasat MM. 2000. Phytoextraction of metals from contaminated soil: A review of plant/soil/metal interaction and assessment of pertinent agronomic issues. *J Hazard Subst Res* 2:1–25.

Lee J, Rai PK, Jeon YJ, Kim KH, Kwon EE. 2017. The role of algae and cyanobacteria in the production and release of odorants in water. *Environ Pollut* 227:252–262.

Leto C, Tuttolonondo T, Bella SL, Leone R, Licata M. 2013. Effect of plant species in a horizontal subsurface flow constructed wetland—Phytoremediation of treated urban wastewater with *Cyperus alternifolius* L. and *Typha latifolia* L. in the west of Sicily (Italy). *Ecol Eng* 61:282–291.

Li HY, Wei DO, Shen M, Zhou ZP. 2012. Endophytes and their role in phytoremediation. *Fungal Divers* 54:11–18.

Li YH, Zhu JN, Liu QF, Liu Y, Liu M, Liu L, Zhang Q. 2013. Comparison of the diversity of root-associated bacteria in *Phragmites australis* and *Typha augustifolia* L. in artificial wetlands. *World J Microbiol Biotechnol* 29:1499–1508.

Lone MA, He Z, Stoffella PJ. 2008. Phytoremediation of heavy metal polluted soils and water: Progresses and perspectives. *J Zhejiang Univ Sci B* 19:210–220.

Loveson A, Sivalingam R, Syamkumar R. 2013. Aquatic macrophyte *Spirodela polyrrhiza* as a phytoremediation tool in polluted wetland water from Eloor, Ernakulam District, Kerala. *J Environ Anal Toxicol* 3(5):1–7.

Lu X, Kruatrachue M, Pokethitiyook P, Homyok K. 2004. Removal of cadmium and zinc by water hyacinth, *Eichhornia crassipes*. *Sci Asia* 30:93–103.

Malar S, Sahi SV, Favas PJC, Venkatachalam P. 2015. Mercury heavy-metal-induced physio-chemical changes and genotoxic alterations in water hyacinths [*Eichhornia crassipes* (Mart.)]. *Environ Sci Pollut Res* 22:4597–4608.

Mani D, Kumar C. 2014. Biotechnological advances in bioremediation of heavy metals contaminated ecosystems: An overview with special reference to phytoremediation. *Int J Environ Sci Technol* 11(3):843–872.

Marques M, Aguiar CRC, Silva JJLS. 2011. Desafios téchnicos e barreiras sociais, econômicas e regulatórias na fitorremediação de solos contaminados. *Rev Bras Ciênc Solo* 35:1–11.

Mazumdar K, Das S. 2015. Phytoremediation of Pb, Zn, Fe, and Mg, with 25 wetland plant species from a paper mill contaminated site in North East India. *Environ Sci Pollut Res* 22:701–710.

Memon AR, Schröder P. 2009. Implications of metal accumulation mechanisms to phytoremediation. *Environ Sci Pollut Res* 16:162–175.

Mishra S, Mohanty M, Pradhan C, Patra HK, Das R, Sahoo S. 2013. Physico-chemical assessment of paper mill effluent and its heavy metal remediation using aquatic macrophytes—A case study at JK paper mill, Rayagada, India. *Environ Monit Assess* 185:4347–4359.

Mishra VK, Tripathi BD. 2008. Concurrent removal and accumulation of heavy metals by the three aquatic macrophytes. *Bioresour Technol* 99:709–712.

Mitsch WJ, Gosselink JG. 2000. *Wetlands*. 3rd ed. New York: John Wiley & Sons.

Mukhopadhyay S, Maiti SK. 2010. Phytoremediation of metal mine waste. *Appl Ecol Environ Res* 8:207–222.

Nahlik AM, Mitsch WJ. 2006. Tropical treatment wetlands dominated by free-floating macrophytes for water quality improvement in Costa Rica. *Ecol Eng* 28:246–257.

Nor YM. 1990. The absorption of metal ions by *Eichhornia crassipes*. *Chem Spec Bioavailab* 2:85–91.

Nunes da Silva M, Mucha AP, Rocha AC, Teixeira C. 2014. A strategy to potentiate Cd phytoremediation by saltmarsh plants—Autochhonous bioaugmentation. *J Environ Manag* 134:136–144.

Nzengya DM, Wishitemi BEL. 2001. The performance of constructed wetlands for wastewater treatment: A case study of Splash wetland in Nairobi, Kenya. *Hydrol Process* 15:3239–3247.

Padmapriya G, Murugesan AG. 2012. Phytoremediation of various heavy metals (Cu, Pb and Hg) from aqueous solution using water hyacinth and its toxicity on plants. *Int J Environ Biol* 2(3):97–103.

Padmavathiamma PK, Li LY. 2007. Phytoremediation technology: Hyper-accumulation metals in plants. *Water Air Pollut* 184:105–126.

Pavel VL, Sobariu DL, Tudorache Fertu ID, Statescu F, Gaverilescu M. 2013. Symbiosis in the environment biomanagement of soils contaminated with heavy metals. *Eur J Sci Theol* 9:211–224.

Philippe AG et al. 2015. Selection of wild macrophytes for use in constructed wetlands for phytoremediation of contaminant mixtures. *J Environ Manag* 147:108–123.

Polechonska L, Klink A. 2014. Trace metal bioindication and phytoremediation potentialities of *Phalaris arundinacea* L. (reed canary grass). *J Geochem Explor* 146:27–33.

Ponce SC, Prado C, Pagano E, Prado FE, Rosa M. 2015. Effect of solution pH on the dynamic of biosorption of Cr (VI) by living plants of *Salvinia minima*. *Ecol Eng* 74:33–41.

Porteous-Moore F, Barac T, Borremans B, Oeyen L, Vangronsveld J, van der Lelie D, Campbell D, Moore ERB. 2006. Endophytic bacterial diversity in polar trees growing on BTEX-contaminated site: The characterisation of isolates with potential to enhance phytoremediation. *Syst Appl Microbiol* 29:539–556.

Preussler KH, Mahler CF, Maranho LT. 2014. Performance of a system of natural wetlands in leachate of a posttreatment landfill. *Int J Environ Sci Technol*. doi: 10.1007/s13762 -014-0674-0.

Rahman KZ, Wiessner A, Kuschk P, van Afferden M, Mattusch J, Müller RA. 2014. Removal and fate of arsenic in the rhizosphere of *Juncus effuses* treating artificial wastewater in laboratory-scale constructed wetlands. *Ecol Eng* 69:93–105.

Rahman MA, Hasegawa H. 2011. Aquatic arsenic: Phytoremediation using floating macrophytes. *Chemosphere* 83:633–646.

Rai PK. 2007a. Phytoremediation of Pb and Ni from industrial effluents using *Lemna minor*: An eco-sustainable approach. *Bull Biosci* 5(1):67–73.

Rai PK. 2007b. Wastewater management through biomass of *Azolla pinnata*: An ecosustainable approach. *Ambio* 36(5):426–428.

Rai PK. 2008a. Phytoremediation of Hg and Cd from industrial effluents using an aquatic free floating macrophyte *Azolla pinnata*. *Int J Phytoremediation* 10(5):430–439.

Rai PK. 2008b. Heavy-metal pollution in aquatic ecosystems and its phytoremediation using wetland plants: An ecosustainable approach. *Int J Phytoremediation* 10(2):133–160.

Rai PK. 2008c. Mercury pollution from chlor-alkali industry in a tropical lake and its biomagnification in aquatic biota: Link between chemical pollution, biomarkers and human health concern. *Human Ecol Risk Assess Int J* 14:1318–1329.

Rai PK. 2009. Heavy metal phytoremediation from aquatic ecosystems with special reference to macrophytes. *Crit Rev Environ Sci Technol* 39(9):697–753.

Rai PK. 2010a. Microcosm investigation on phytoremediation of Cr using *Azolla pinnata*. *Int J Phytoremediation* 12:96–104.

Rai PK. 2010b. Phytoremediation of heavy metals in a tropical impoundment of industrial region. *Environ Monit Assess* 165:529–537.

Rai PK. 2010c. Seasonal monitoring of heavy metals and physico-chemical characteristics in a lentic ecosystem of sub-tropical industrial region, India. *Environ Monit Assess* 165:407–433.

Rai PK. 2010d. Heavy metal pollution in lentic ecosystem of sub-tropical industrial region and its phytoremediation. *Int J Phytoremediation* 12(3):226–242.

Rai PK. 2011. *Heavy Metal Pollution and Its Phytoremediation through Wetland Plants*. New York: Nova Science Publisher.

Rai PK. 2012. Assessment of multifaceted environmental issues and model development of an Indo-Burma hot spot region. *Environ Monit Assess* 184:113–131.

Rai PK. 2013. *Plant Invasion Ecology: Impacts and Sustainable Management*. New York: Nova Science Publisher.

Rai PK. 2015a. Paradigm of plant invasion: Multifaceted review on sustainable management. *Environ Monit Assess* 187:759–785.

Rai PK. 2015b. What makes the plant invasion possible? Paradigm of invasion mechanisms, theories and attributes. *Environ Skeptics Critics* 4(2):36–66.

Rai PK. 2015c. Concept of plant invasion ecology as prime factor for biodiversity crisis: Introductory review. *Int Res J Environ Sci* 4(5):85–90.

Rai PK. 2015d. *Environmental Issues and Sustainable Development of North East India.* Saarbrücken, Germany: Lambert Academic Publisher.

Rai PK. 2016. Biodiversity of roadside plants and their response to air pollution in an Indo-Burma hotspot region: Implications for urban ecosystem restoration. *J Asia Pac Biodivers* 9:47–55.

Rai PK et al. 2018. A critical review of ferrate(VI)-based remediation of soil and groundwater. *J Environ Res* 160:420–448.

Rai PK, Chutia B. 2016. Biomagnetic monitoring through *Lantana* leaves in an Indo-Burma hot spot region. *Environ Skeptics Critics* 5(1):1–11.

Rai PK, Chutia BM, Patil SK. 2014. Monitoring of spatial variations of particulate matter (PM) pollution through bio-magnetic aspects of roadside plant leaves in an Indo-Burma hot spot region. *Urban Forest Urban Green* 13:761–770.

Rai PK, Mishra A, Tripathi BD. 2010. Heavy metals and microbial pollution of river Ganga: A case study on water quality at Varanasi. *Aquat Ecosyst Health Manag* 13(4):352–361.

Rai PK, Panda LS, Chutia BM, Patil SK. 2013. Bio-monitoring of particulates through magnetic properties of road-side plant leaves: A case study of Aizawl, Mizoram, India. *Sci Technol J* 1(1):31–35.

Rai PK, Singh M. 2016. *Eichhornia crassipes* as a potential phytoremediation agent and an important bioresource for Asia Pacific region. *Environ Skeptics Critics* 5(1):12–19.

Rai PK, Singh MM. 2015. *Lantana camara* invasion in urban forests of an Indo-Burma hotspot region and its ecosustainable management implication through biomonitoring of particulate matter. *J Asia Pac Biodivers* 8:375–381.

Rai PK, Tripathi BD. 2009. Comparative assessment of *Azolla pinnata* and *Vallisneria spiralis* in Hg removal from G.B. Pant Sagar of Singrauli Industrial region, India. *Environ Monitor Assess* 148:75–84.

Raghukumar, C. 2008. Marine fungal biotechnology: An ecological perspective. *Fungal Diversity* 31:19–35.

Ranieri E, Fratino U, Petruzzelli D, Borges AC. 2013. A comparison between *Phragmites australis* and *Helianthus annuus* in chromium phytoextraction. *Water Air Soil Pollut* 224:1465.

Sachdeva S, Sharma A. 2012. *Azolla*: Role in phytoremediation of heavy metals. In *Proceedings of the National Conference "Science in Media 2012,"* pp. 9–14.

Salt DE, Blaylock M, Nanda Kumar PBA, Dushenkhov V, Ensley BD, Raskin I. 1995. Phytoremediation: A novel strategy for the removal of toxic metals from the environment using plants. *Biotechnology* 13:468–474.

Salt DE, Pickering IJ, Price RC, Gleba D, Dushenkhov S, Smith RD, Raskin I. 1997. Metal accumulation by aqua-cultured seedlings of Indian mustard. *Environ Sci Technol* 31(6):1636–1644.

Salt DE, Smith RD, Raskin I. 1998. Phytoremediation. *Annu Rev Plant Physiol Plant Mol Biol* 49:643–648.

Sasidharan NK, Azim T, Devi DA, Mathew S. 2013. Water hyacinth for heavy metal scavenging and utilization as organic manure. *Indian J Weed Sci* 45(3):204–209.

Sharma P. 2007. Material translocation characteristics and the effect of soil nutrient on the growth of *Typha austifolia*. Dissertation, University of Saitama, Saitama, Japan.

Sharma S, Singh S, Manchanda VK. 2014. Phytoremediation: Role of terrestrial plants and aquatic plants and aquatic macrophytes in the remediation of radionuclides and heavy metal contaminated soil and water. *Environ Sci Pollut Res*. doi: 10.1007/s11356-014-3635-8.

Sharma SS, Gaur JP. 1995. Potential of *Lemna polyrrhiza* for removal of heavy metals. *Ecol Eng* 4:37–43.

Shehzadi M, Fatima K, Imran A, Mirza MS, Khan QM, Afzal M. 2016. Ecology of bacterial endophytes associated with wetland plants growing in textile effluent for pollutant-degradation and plant growth-promotion potentials. *Plant Biosyst* 150:1261–1270.

Shutes RBE. 2001. Artificial wetlands and water quality improvement. *Environ Int* 26:441–447.

Singh MM, Rai PK. 2016. Microcosm investigation of Fe (iron) removal using macrophytes of Ramsar Lake: A phytoremediation approach. *Int J Phytoremediation* 18(12):1231–1236.

Soda S, Hamada Y, Yamaoka Y, Ike M, Nakazato H, Saeki Y, Kasamatsu T, Sakurai Y. 2012. Constructed wetlands for advanced treatment of wastewater with a complex matrix from a metal-processing plant: Bioconcentration and translocation factors of various metals in *Acorus gramineus* and *Cyperus alternifolius*. *Ecol Eng* 39:63–70.

Sooksawat N, Meetam M, Kruatrachue M, Pokethitiyook P, Nathalang K. 2013. Phytoremediation potential of charophytes: Bioaccumulation and toxicity studies of cadmium, lead and zinc. *J Environ Sci (China)* 25(3):596–604.

Souza FA, Dziedzic M, Cubas SA, Maranho LT. 2013. Restoration of polluted waters by phytoremediation using *Myriophyllum aquaticum* (Vell.) Verdc. Haloragaceae. *J Environ Manag* 120:5–9.

Stone KC, Poach ME, Hunt PG, Reddy GB. 2004. Marsh-pond-marsh constructed wetland design analysis for swine lagoon wastewater treatment. *Ecol Eng* 23:127–133.

Suelee L et al. 2017. Phytoremediation potential of vetiver grass (*Vetiveria zizanioides*) for treatment of metal-contaminated water. *Water Air Soil Pollut* 228:158.

Susarla S, Medina VF, McCutcheon SC. 2002. Phytoremediation: An ecological solution to organic chemical contamination. *Ecol Eng* 18:647–658.

Tiwari S, Dixit S, Verma N. 2007. An effective means of biofiltration of heavy metal contaminated water bodies using aquatic weed *Echhornia crassipes*. *Environ Monit Assess* 129:253–256.

Turker OC, Bocuk H, Yakar A. 2013. The phytoremediation ability of a polyculture constructed wetland to treat boron from mine effluent. *J Hazard Mater* 252–253:132–141.

Ullah A, Mushtaq H, Ali H, Munis MFH, Javed MT, Chaudhary HJ. 2014. Diazotrophs-assisted phytoremediation of heavy metals: A novel approach. *Environ Sci Pollut Res.* doi: 10.1007/s11356-014-3699-5.

Upadhyay AK, Singh NK, Rao UN. 2014. Comparative metal accumulation potential of *Potamogeton pectinus* L. and *Potamogeton crispus* L.: Role of enzymatic and non-enzymatic antioxidants in tolerance and detoxification of metals. *Aquat Bot* 117:27–32.

Verhoeven JTA, Meuleman AFM. 1999. Wetlands for wastewater treatment: Opportunities and limitations. *Ecol Eng* 12(1):5–12.

Wu J, Yang L, Zhong F, Cheng S. 2014. A field study on phytoremediation of dredged sediment contaminated by heavy metals and nutrients: The impacts of sediment aeration. *Environ Sci Pollut Res* 21:13452–13460.

Xue PY, Li GX, Liu WJ, Yan CZ. 2010. Copper uptake and translocation in a submerged aquatic plant *Hydrilla verticillata* (L.f.) Royle. *Chemosphere* 81:1098–1103.

Yadav AK, Abbassi R, Kumar N, Satya S, Sreekrishnan TR, Mishra BK. 2012. The removal of heavy metals in wetland microcosms: Effects of bed depth, plant species, and metal mobility. *Chem Eng J* 211–212:201–507.

Yadav R, Arora P, Kumar S, Chaudhury A. 2010. Perspectives for genetic engineering of poplars for enhanced phytoremediation abilities. *Ecotoxicology* 19:1574–1588.

Yang X, Chen S, Zhang R. 2014. Utilization of two invasive free-floating aquatic plants (*Pistia stratiotes* and *Eichhornia crassipes*) as sorbents for oil removal. *Environ Sci Pollut Res* 21:781–786.

Ye Z, Wong M, Lan C. 2004. Use of a wetland system for treating Pb/Zn mine effluent: A case study in southern China from 1984 to 2002. In *Wetland Ecosystems in Asia: Function and Management*, ed. M Wong, 413–434. Amsterdam: Elsevier.

Yuan X, Huang H, Zeng G, Li H, Wang J, Zhou C, Zhu H, Pei X, Liu Z, Liu Z. 2011. Total concentrations and chemical speciation of heavy metals in liquefaction residues of sewage sludge. *Bioresour Technol* 102(5):4104–4110.

Zhang Y, He L, Chen Z, Zhang W, Wang Q, Qian M, Sheng X. 2011. Characterization of lead-resistant and ACC deaminase-producing endophytic bacteria and their potential in promoting lead accumulation of rape. *J Hazard Mater* 186:1720–1725.

Zhu YL, Zayed AM, Quian JH, De Souza M, Terry N. 1999. Phytoaccumulation of trace elements by wetland plants. II. Water hyacinth. *J Environ Qual* 28:339–344.

3 Progress, Prospects, and Challenges of Phytoremediation with Wetland Plants

INTRODUCTION

Freshwater resources are extremely relevant for the sustainable development of people in the current Anthropocene era. Being a major source of freshwater, the need for the conservation of wetlands from contamination is important. It is the most productive ecosystem on the earth (Ghermandi et al., 2008). Wetland is defined as land that is transitional between terrestrial and aquatic systems, where the table is usually at or near the surface, or the land is periodically covered with shallow water, and which under normal circumstance supports vegetation typically adapted to life in water-saturated soil (Oberholster et al., 2014). Wetlands are categorized into marine (coastal wetlands), estuarine (including deltas, tidal marshes, and mangrove swamps), lacustarine (lakes), riverine (rivers and streams), and palustarine ("marshy"—marshes, swamps, and bogs) based on their hydrological, ecological, and geological characteristics (Cowardin et al., 1979; Rai, 2008b, 2009a; Bassi et al., 2014; Singh and Rai, 2016).

In the current scenario, water in most Ramsar wetlands has been heavily degraded, mainly due to agricultural runoff of pesticides and fertilizers, and industrial and municipal wastewater discharge, all of which cause widespread eutrophication and algal bloom and a resulting odor (Roy, 1992; Devi and Sharma, 2002; Prasad et al., 2002; Sanjit et al., 2005; Liu and Diamond, 2005; Umavathi and Longankumar, 2010; Singh et al., 2010; LDA i.e. Loktak Development Authority, 2011; Rai and Raleng, 2011; Singh et al., 2014; Singh and Rai, 2016; Lee et al., 2017). Heavy metal and other emerging contaminant pollution in aquatic and wetland systems has become an important point of discussion in recent years (Rai, 2007a,b, 2008a–c, 2009a, 2010a–d, 2011; Rai et al., 2010; Singh and Rai, 2016). Sources of emerging contaminants like heavy metals are coal mining and its allied industries, for example, thermal power plants, and also chemical industries, for example, chlor-alkali plants in developing countries like India. Effluents from these industries pose serious threats to water quality and the aquatic biodiversity of rivers, lakes, and reservoirs (Pip and Stepaniuk, 1992; Rai, 2011, 2012). The problem of emerging contaminants and heavy metal pollution is materializing as a matter of concern at the local, regional, and global scale. Heavy metal and emerging contaminant pollution in the aquatic ecosystem pose a serious threat to aquatic biodiversity, and drinking contaminated water poses severe health hazards for humans. Therefore, attention has been shifted from the mere monitoring of environmental conditions to the development of alternative means to solve environmental problems at local and global levels. The economic aspects and side effects of conventional treatment technologies in aquatic ecosystems have paved the way to phytoremediation technology. In phytoremediation, plants are used to ameliorate the environment from various hazardous pollutants. It is a cost-effective and eco-friendly technology for environmental cleanup (Rai, 2011).

IMPORTANCE OF WETLANDS IN THE CURRENT ANTHROPOCENE

Wetlands possess significantly high ecosystem service value (Costanza et al., 1997; Zhang et al., 2013). Wetlands have long been providing fisheries, irrigation and domestic water supply, recreation, and tourism (Jain et al., 2007). Wetlands are the most productive ecosystem on the earth (Ghermandi et al., 2008). They are noted for contributing groundwater recharge, supporting a rich diversity of aquatic flora and fauna (Bassi et al., 2014), improving water quality, abating flood waters, supporting biodiversity, and storing carbon (Moreno-Mateos et al., 2012; Doherty et al., 2014). Wetlands, which include peatlands, mangroves, mires, marshes, and swamps, help in the carbon cycle (Bassi et al., 2014). Wetland sediments store carbon in the long term, while the wetland biomass of plants, animals, bacteria, fungi, and dissolved components in surface and ground water store it in the short term (Wylynko, 1999). Natural wetlands have been used for centuries as a sink for waste, being capable of assimilating large amounts of environmental pollutants (Sheoran and Sheoran, 2006). Wetlands act as a low-cost measure to reduce point and nonpoint-source pollution (Bystrom et al., 2000). Wetlands also act as the "kidney of nature," which has the capability to trap and/or efficiently modify a broad spectrum of contaminants (Mander and Mitsch, 2009). These ecosystems have special characteristics that make them particularly suitable for wastewater treatment: they are semi-aquatic systems that normally contain large quantities of water, they have oxic and partly anoxic soils where the biodegradation of organic matter takes place, and they support highly productive, tall emergent vegetation capable of taking up a large amount of nutrients that enhance growth (Verhoeven and Meuleman, 1999; Arroyo et al., 2013). They support wildlife, which may include many rare, threatened, and endangered species (EPA, 2004), and perform various ecological functions.

Wetlands are one of the fastest deteriorating ecosystems due to urbanization, which has led to the contribution of increasing pollution and contamination. The introduction of relatively recent anthropogenic originated toxic substances, including heavy metals, and their massive relocation to different environmental compartments, especially water, has resulted in severe pressure on the self-cleansing capacity of aquatic ecosystems (Rai, 2009a). Heavy metals have a serious deleterious effect on living organisms, especially humans. Attaining sustainability and following the green approach in pollution remediation, phytotechnology is the need of the hour (Rai, 2012).

HEAVY METALS: AN EMERGING CONTAMINANT OF GLOBAL CONCERN

In Chapter 1, we discussed various emerging contaminants except metals; the focus of this chapter is oriented toward this pollutant of global concern. Heavy metals are metals of the d-block, having a specific gravity greater than 5 g/cm^3 (Nies, 1999; Rai, 2012). Increasing industrialization and urbanization have caused the problem of heavy metals to increase. They are listed as priority pollutants by the U.S. Environmental Protection Agency. There are more than 70,000 chemicals in use in the world, and more than 4 million on the American Chemical Society's computer list of chemicals (Cairns et al., 1988; Rai, 2009a, 2012). As hazards lead, mercury, arsenic, and cadmium are ranked first, second, third, and sixth, respectively, on the list of the U.S. Agency for Toxic Substances and Disease Registry (ATSDR), which lists all hazards present in toxic waste sites according to their prevalence and the severity of their toxicity (Rai, 2009a).

Pollution of the biosphere with toxic metals has accelerated dramatically since the beginning of the Industrial Revolution (Nriagu, 1979). The problem of heavy metal pollution is emerging as a matter of concern at the local, regional, and global scale. In Japan, 2252 people were affected and 1043 died from Minamata disease over the past few decades, caused by elevated mercury pollution from a chemical plant (Kudo and Miyahara, 1991; Rai, 2011, 2012). Recently, in developing countries major problems have arisen, such as groundwater contamination with As in Bangladesh (Alam et al., 2003; Rai, 2009a, 2012) and heavy metal contamination, for example, Cd, Pb, Cu, and Zn in drinking water sources in Bolivia, Hong Kong, and Berlin (Ho et al., 2003; Zietz et al., 2003; Miller et al., 2004). Heavy metals in surface water systems can be from natural or anthropogenic sources. Currently, anthropogenic inputs of metals exceed natural inputs. High levels of Cd, Cu, Pb, and Fe can act as ecological toxins in aquatic and terrestrial ecosystems (Balsberg-Pahlsson, 1989; Guilizzoni, 1991; Rai, 2012). Excess metal levels in surface water may pose a health risk to humans and the environment.

GLOBAL SOURCES OF HEAVY METALS AND OTHER EMERGING CONTAMINANTS

Heavy metals or other emerging contaminants in essential and nonessential forms are naturally persistent in the environment (Alhashemi et al., 2011). The inorganic pollutants and emerging contaminants present in aquatic bodies originate from natural and anthropogenic sources (Mdegela et al., 2009), but the occurrence of heavy metals and emerging contaminants in the environment is mainly due to anthropogenic sources (Zhipeng et al., 2009). The point source discharges of municipal sewage, industrial facility effluents, and nonpoint-source discharges from domestic waste, fisheries, and agriculture contaminate aquatic bodies with toxic heavy metals or other emerging contaminants (Yuwono et al., 2007). Agro-chemicals, including fertilizer and plant nutrients, also lead to increases in the concentrations of heavy metals and emerging contaminants in water and soil (Rattan et al., 2005). This increased contamination of aquatic bodies is leading to a serious threat to public health from the consumption of fish from these sources (Bickham et al., 2000; Mayon et al., 2006; Fatima et al., 2014). The toxicity of emerging contaminants and metals is a highly influential geochemical factor in metal bioavailability (Fairbrother et al., 2007). The sediments of the wetland effectively sequester hydrophobic chemical pollutants, which are readily available from various pollutant discharges

(Harikumar et al., 2009). The discharges of specific local sources from dye formulators and paint (Cd, Cr, Cu, Hg, Pb, Se, and Zn), metal-based industries (Cd, Cr, and Zn from electroplating), smelters (Cu, Ni, and Pb), petroleum refineries (As and Pb), and chemical factories may lead to metal and emerging contaminant accumulation in sediments (Al-Masri et al., 2002; Harikumar et al., 2009).

ANTHROPOGENIC SOURCES OF HEAVY METALS AND EMERGING CONTAMINANTS

Agriculture seems to be the most considerable source of pollution from emerging contaminants due to runoff from fertilized land (Rai and Tripathi, 2008; Rai, 2009a, 2011, 2012). Industrial processing and solid waste dumps are considered to be the main anthropogenic sources of emerging contaminants and metal pollution. As a result of mining and metal working in ancient times, a close link has been demonstrated between metal pollution and human history (Nriagu, 1996; Rai, 2011, 2012). The primary sources of emerging contaminants and metal pollution are the burning of fossil fuels, mining and smelting of metalliferous ores, municipal wastes, fertilizers, pesticides, and sewage (Rai, 2007a,b, 2008a–c, 2009a,b, 2010a–d; Rai and Tripathi, 2009; Rai et al., 2010). Emerging contaminants or heavy metal contamination and acid mine drainage (AMD) are very important concerns where waste materials containing metal-rich sulfides from mining activity have been stored or abandoned (Concas et al., 2006; Rai, 2009a). Further sources of emerging contaminants are coal mining (Finkelman and Gross, 1999) and its allied industries, for example, thermal power plants, and also chemical industries, for example, chloralkali plants, are major sources of emerging contaminants and heavy metals in industrial belts in developing countries like India (Sharma, 2003; Rai et al., 2007). Effluents of these industries pose serious threats to water quality and the aquatic biodiversity of rivers, lakes, and reservoirs.

Major sources of emerging contaminants, like mercury, include gold (Au) and silver (Ag) mines, the coal industry, untreated discarded batteries, and industrial waste disposal (Pilon-Smits and Pilon, 2002; Malar et al., 2015). Emerging contaminants like Cd are used in a number of industries, such as electronic components, electroplating, metal mechanic processing, and mining (Haddam et al., 2011). They can also be found in plastic toys and food containers (Kumar and Pastore, 2007; Liu et al., 2014). Anthropogenic sources of environmental manganese include sewage sludge, mining and mineral processing, and combustion of fossil fuels (WHO, 2004). The release of these toxic emerging contaminants in biologically available forms by human activity may damage or alter both natural and man-made ecosystems (Chatterjee et al., 2011).

EMERGING CONTAMINANTS AND HEAVY METALS AS POLLUTANTS TO WETLANDS AND THEIR ENVIRONMENT

Heavy metals and emerging contaminants are becoming one of the major threats and important issues for wetland environments. Many emerging contaminants that are potential toxic materials and environmental pollutants have drawn great attention from around the world (Liu et al., 2009; Zhang et al., 2012; Montuori et al., 2013; Zhang et al., 2014). The aquatic ecosystem contaminated from emerging contaminants and heavy metals has been an urgent problem worldwide (Alhashemi et al., 2011). These emerging contaminants and heavy metals include aluminum (Al), arsenic (As), cadmium (Cd), copper (Cu), cobalt (Co), chromium (Cr), mercury (Hg), iron (Fe), lead (Pb), magnesium (Mg), manganese (Mn), nickel (Ni), strontium (Sr), vanadium (V), and zinc (Zn). These heavy metals are persistent toxic substances and the leading cause of aquatic contamination (Gurcu et al., 2010; Fatima et al., 2014). Heavy metals and emerging contaminants can cause toxicity in biological systems through bioaccumulation, and can also cause a hazardous effect in the wetlands environment and associated living organisms through biomagnification (increase of toxic accumulation in organisms through various food chain systems). Their toxicity has been of great concern since they are very important to the health of people and ecology (Feng et al., 2008). The occurrence of toxic emerging contaminants in lake, reservoir, and river water affects the lives of local people that depend on these water sources for their daily requirements (Rai et al., 2002).

Wetlands have long been recognized as an important emerging contaminant and heavy metal sink due to their chemical, physical, and biological processes involving adsorption, precipitation, sedimentation, setting, and induced biogeochemical changes by plants and bacteria (Chandra et al., 2013; Jiao et al., 2014; Xin et al., 2014). The wetlands are being degraded by human activities and the discharge of emerging contaminants, such as persistent organic pollutants (Dsikowitzky et al., 2011) and heavy metals (Noegrohati, 2005). The persistent emerging contaminants, like some heavy metal tissue bioaccumulation, due to their redox cycling and their ability to deteriorate plant tissue sulfahydryl groups, lead to oxidative stress, which can be seen through genotoxic parameters such as the micronucleus text (MNT) and comet assay (Ali et al., 2009). An excessive concentration of emerging contaminants in soil solutions, such as Co, Cu, Fe, Mn, Mo, Ni, and Zn, limits the plant enzyme and protein functions, metabolism, growth, and development (Mengel at al., 2001; Hänsch and Mendel, 2009). Many microorganisms have genetic variability that allows them to circumvent the toxic effect caused by the emerging contaminants, but the more developed organisms usually succumb to the toxic effects (Valdman et al., 2001).

HEALTH IMPACT OF EMERGING CONTAMINANTS AND HEAVY METAL POLLUTION

Emerging contaminants and their pollution of surface and ground waters, as well as air (specifically magnetic particles of particulate matter), pose a major environmental and health problem that is still in need of an effective and affordable technological solution (Baldantoni et al., 2004, 2005; Pandey et al., 2005; Rai, 2010a–d, 2013, 2016a; Rai et al., 2010, 2013, 2014, 2018; Rai and Panda, 2014; Lee et al., 2017). Exposure of the population to emerging contaminants and heavy metals may cause neurobehavioral disorders, such as fatigue, insomnia, decreased concentration, depression, irritability, and sensory and motor symptoms (Hanninen and Lindstrom, 1979). Exposure to emerging contaminants and heavy metals has also been linked to developmental retardation, various types of cancers, kidney damage, autoimmunity, and even death in some instances of exposure to very high concentrations (Glover-Kerkvilet, 1995). More specifically, deleterious emerging contaminants, such as methylmercury intake through fish and aquatic foods, can have a considerable effect on human health. At higher levels, mercury can damage vital organs, such as the lungs and kidneys. Methylmercury may cross the placental barrier and cause fetal brain damage (Sharma, 2003; Rai, 2009a). Accumulation of emerging contaminants like Cd in the human body (principally in the kidney and liver) can cause renal dysfunction and bone disease, such as itai-itai in Japan (Nordberg, 1996). High concentrations of cadmium have been reported in the sewage, irrigation water, and vegetables grown in the Gangetic Plain of the eastern Uttar Pradesh and western Bihar regions of India, resulting in gallbladder carcinoma and the production of stones, with the cadmium, chromium, and lead concentrations being significantly higher in gallbladder carcinoma than in gallstones (Shukla et al., 1998; Rai and Tripathi, 2008). Lead poisoning in children causes neurological damage, leading to reduced intelligence, loss of short-term memory, learning disabilities, and coordination problems. The effects of arsenic include cardiovascular problems, skin cancer and other skin effects, peripheral neuropathy, and kidney damage (WHO, 1997).

Numerous human health problems, such as developmental, reproductive, cardiovascular, gastrointestinal, dermal, immunological, hepatic, hematological, renal, neurological, respiratory, genotoxic, mutagenic, and carcinogenic effects (such as liver cancer) (Lin et al., 2013; Rai, 2015e, 2016a; Singh and Prasad, 2015), have been caused by emerging contaminants like heavy metals. Toxic heavy metals such as Pb, Cr, and As can cause a serious problems to animals and humans, such as irritation of sensory organs, respiratory troubles, and arsenic poisoning (Yu et al., 2014). Drinking water containing emerging contaminants, for example, As and Cd, can cause cancer (Steinemann, 2000), allergies, and hyperpigmentation (Wongsasuluk et al., 2014), and may also seriously damage the kidney, liver, digestive system, nervous system (Wcislo et al., 2002; Li et al., 2008), lungs, brain, and bones (PCD, 2000). Chronic exposure to low doses of cancer-causing heavy metals and emerging contaminants may induce many types of cancer (Singh and Prasad, 2015). Cd, an emerging contaminant, is carcinogenic under inorganic heavy metal (Lauwerys, 1979) and can reside inside the body for several years, as it has a half-life of 38 years (Berman, 1980). Cd can infiltrate into organisms through the respiratory tract and skin and then damage tissues and organs owing to its interactions with biological macromolecules (Nursita et al., 2009). Pb also affects health, causing cardiovascular, nervous system, blood, and bone diseases (Jarup, 2003).

Another reason for toxic emerging contaminants like heavy metals causing concern is that they may be transferred and accumulated in the bodies of animals or humans through the food chain, which would probably cause DNA damage and carcinogenic effects due to their mutagenic ability (Knasmuller et al., 1998; Rai, 2008b, 2009a). For example, some species of Cd, Cr, and Cu have been associated with health effects, ranging from dermatitis to various types of cancer (Das et al., 2014; McLaughlin et al., 1999). In addition, some metals occur in the environment as radioactive isotopes (e.g., 238U, 137Cs, 239Pu, and 90Sr), which can greatly increase the health risk (Pilon-Smits and Pilon, 2002; Rai et al., 2010; Rai, 2015e).

TABLE 3.1
Capital and Operational Costs for Chemical Management in Industries

S. No.	Chemical Technology	Capital Costs (US$/m³)		Operating Costs (US$/m³)
		A	B	
1	Precipitation (includes neutralization, coagulation, flocculation, and separation)	12.5	8	0.003–0.013
2	Adsorption (through granulated activated carbon)	500	250	0.020–0.050
3	Membrane filtration	12.5	11	0.013–0.050
4	Ion exchange	100	75	0.050–0.250

Source: Modified after Eccles, 1999; Rai, P.K., *Heavy Metal Pollution and Its Phytoremediation through Wetland Plants*, Nova Science Publisher, New York, 2011; Rai, P.K., *Environ. Monit. Assess.*, 184, 113–131, 2012.

Note: A is for installed industrial process potential of 1000 m³ d⁻¹. Whereas, B is for installed industrial process potential of 10000–20000 m³ d⁻¹.

HEAVY METALS AND SOURCE MANAGEMENT

The chemical management methodology with reference to heavy metals should be:

- Compatible
- Cost-effective
- Flexible enough to handle fluctuations in the quality and quantity of effluent feed
- Reliable, by operating continuously
- Robust enough to minimize supervision and maintenance
- Selective enough to remove only the contaminant metals under consideration
- Simple enough to minimize automation and the need for skilled operators (Eccles, 1999; Rai, 2012) (Table 3.1)

ADVANTAGES AND DISADVANTAGES OR LIMITATIONS OF PHYTOREMEDIATION

Phytoremediation technology—and its application in wetlands for contamination treatment—is a productive method that has many advantages and disadvantages (Farraji et al., 2016). Some of the advantages are given below (see Table 3.2 for detail).

- It has been adopted by many researchers and scientists because of its low cost, effectiveness, easy applicability to a wide range of water contaminants, free plants for use in the method, safety, and eco-friendly nature.
- It is a low-cost option for environmental media, and it is suited to large sites that have relatively low contamination (Ginneken et al., 2007).
- Phytoremediation can be performed aesthetically with a clean environment.

TABLE 3.2

Summary of the Treatability of Physicochemical Treatments for Industrial Effluents (Types of Treatment, Advantages, and Disadvantages)

Number	Conventional Method	Advantages	Disadvantages
1	Chemical precipitation	Low capital cost, simple operation	Sludge generation, extra operational cost for sludge disposal
2	Coagulation–flocculation	Shorter time to settle out suspended solids, improved sludge settling	Sludge production, extra operational cost for sludge disposal
3	Dissolved air flotation	Low cost, shorter hydraulic retention time	Subsequent treatments are required to improve the removal efficiency of heavy metals
4	Ion exchange	No sludge generation, less time-consuming	Not all ion exchange resin is suitable for metal removal, high capital cost
5	Ultrafiltration	Smaller space requirement	High operational cost, prone to membrane fouling
6	Nanofiltration	Lower pressure than reverse osmosis (7–30 bar)	Costly, prone to membrane fouling
7	Reverse osmosis	High rejection rate, able to withstand high temperatures	High energy consumption due to high pressure required (20–100 bar), susceptible to membrane fouling

Source: Rai, P.K., *Crit. Rev. Environ. Sci. Technol.*, 39(9), 697–753, 2009; Rai, P.K., *Heavy Metal Pollution and Its Phytoremediation through Wetland Plants*, Nova Science Publisher, New York, 2011.

- It is also an eco-friendly technology and can use freely available macrophytes for the sustainability of the process.
- The biomass production from the technology can also be used in different forms. The accumulated metals can be obtained from the biomass after burning into ash.
- Phytoremediation can be performed for the treatment of effluents and wastewater from agriculture, industry, municipalities, storm water, and so forth.

There are also some disadvantages or limitations in phytoremediation.

- It is a time-consuming process and is totally dependent on the plant deployed for the treatment.
- Plant growth takes time, and the growth is due to the availability of the proper medium.
- During the process of phytoremediation, plants die for many reasons, such as contamination extremity, lack of nutrients, and lack of photosynthesis. Due to a high concentration of contaminant, the toxicity level increases, which leads to the death of the plants.

- Appropriate plant selection is required for an effective phytoremediation process. The growth rate of the mature plant is slow and attends withering processes. Plants with a high growth rate and resistant to environmental stress are more suitable for phytoremediation. Each plant has the capacity to remove different heavy metals and contaminants according to its metabolic activities.
- The timely removal of these plants from the site is required.
- Climatic conditions are one of the important factors supporting the phytoremediation process.

SWOT ANALYSIS FOR PHYTOTECHNOLOGIES AND PHYTOREMEDIATION

A strengths, weaknesses, opportunities, and threats (SWOT) analysis is necessary to fill the knowledge gap in the context of phytotechnologies to direct focused future research (Gomes, 2012).

Strengths: The strength of phytotechnologies is that they can be an eco-friendly, cost-effective (thus having higher public acceptance), and solar-driven way of functioning. Moreover, environmental benefits, such as soil erosion control and carbon sequestration, and socioeconomic benefits add to the strengths (in the context of a SWOT analysis) of phytoremediation.

Weaknesses: As stated earlier, a phytotechnology tool should be fast growing, have high biomass and deep roots, be easy to harvest, and tolerate and accumulate a range of emerging contaminants in its above- and below-ground biomass; thus, screening potent plants with such attributes is a prerequisite for the success of phytotechnologies, and this feature forms the weaknesses component of the SWOT analysis for phytoremediation. Further, the slow pace of phytotechnologies, their specificity to only certain emerging contaminants, and the adaptability of plants to particular tropical or temperate regions can be considered other under weaknesses.

Opportunities: The opportunities component of the SWOT analysis lies in the coupling of phytotechnologies with other eco-technologies for bioenergy, phytomining, and so forth, to achieve economic returns from phytotechnologies, nanoparticle (NP) generation for diverse applications in manufacturing and the health sectors, the prospects of advanced instrumentation and genetic engineering to enhance the pace of phytoremediation, and applications in the designer approach of constructed wetlands to degrade diverse emerging contaminants.

Threats: Threats may be the biomass utility issue after phytoremediation, their possible leakage into the total environment, and the invasive nature of certain wetland plants, which in my opinion is addressed satisfactorily and can be considered an opportunity. All these issues are discussed elsewhere in this book.

A list of wetland plants used for the eco-removal of metals (as emerging contaminants of concern) in phytoremediation by various researchers and scientists is shown in Table 3.3.

TABLE 3.3
Wetland Plants and Accumulated Metals (as Emerging Contaminants of Concern) in Phytoremediation

Scientific Names of Plants	Emerging Contaminant Names	Concentrated Area	References
Duckweeds (*Lemna, Spirodela, Wolffia, Wolffiella,* and *Landoltia*)	Emerging contaminants (pharmaceuticals, personal care products, pesticides, and surfactants)	Root, shoot/whole plant	Forni and Tommasi, 2015
Phragmites karka and *Vetiveria nigritana*	Emerging contaminants	Root, shoot/whole plant	Badejo et al., 2015
Eichhornia crassipes	Pb, Cr, Zn, Mn, and Cu	Water, sediment, and whole plants	Tiwari et al., 2007; Singh and Rai, 2016
Polygonum sp. and *Ludwigia* sp.	Fe, Mn, Ni, Cu, Zn, As, and Pb	Sediment and whole plants	Bako and Daudu, 2007
Typha latifolia	Mn, Cu, Cd, Co, Zn, Pb, Ni, and Cr	Water, sediment, and leaves and roots of the plants	Sasmaz et al., 2008
Ipomoea aquatica, Eichhornia crassipes, Typha angustata, Echinochloa colonum, Hydrilla verticillata, Nelumbo nucifera, and *Vallisneria spiralis*	Cd, Co, Cu, Ni, Pb, and Zn	Water, sediment, and leaves, stems, and roots of the plants	Kumar et al., 2008; Singh and Rai, 2016
Azolla pinnata and *Vallisneria spiralis*	Hg	Water, sediment, and whole plants	Rai and Tripathi, 2009; Singh and Rai, 2016
Aponogeton natans, Cyperus rotundus, Hydrilla verticillata, Ipomoea aquatica, Marsilea quadrifolia, Potamogeton pecitnatus, Eichhornia crassipes, Lemna minor, Spirodela polyrhiza, Azolla pinnata, Vallisneria spiralis, and *Polygonum amphibium*	Cu, Cr, Fe, Pb, Zn, Hg, Ni, Cd, and Mn	Effluent, water, sediment, and whole plants	Rai, 2009b
Eichhornia crassipes, Lemna minor, and *Azolla pinnata*	Cu, Cr, Fe, Mn, Ni, Pb, Zn, Hg, and Cd	Effluent, water, sediment, and whole plants	Rai, 2010; Singh and Rai, 2016

(Continued)

TABLE 3.3 (CONTINUED)
Wetland Plants and Accumulated Metals (as Emerging Contaminants of Concern) in Phytoremediation

Scientific Names of Plants	Emerging Contaminant Names	Concentrated Area	References
Hydrilla verticillata, Elodea canadensis, and Salvinia sp.	Cu, Fe, and Ni	Water	Begam and HariKrishna, 2010; Singh and Rai, 2016
Eichhornia crassipes, Ipomoea aquatica, Wolffia arrhiza, Pistia stratiotes, Trapa bispinosa, Cynodon dactylon, Scirpus sp., Cyperus rotundus, Colocasia esculenta, and Sagittaria montevidensis	Ca, Fe, Mn, Cr, Zn, Cu, and Pb	Marginal bank soils, water, sediment, and whole plants	Chatterjee et al., 2011; Singh and Rai, 2016
Schoenoplectus californicus and Ricciocarpus natans	Zn, Cu, and Cd	Sediment and whole plants	Boudet et al., 2011
Bidens tripartitus, Polygonum amphibium, Lycopus europaeus, Typha angustifolia, and Typha latifolia	Fe, Mn, Cu, and Pb	Water, sediments, and leaves, stems, and roots of the plants	Branković et al., 2011
Ceratophyllum demersum, Echinochloa pyramidalis, Echhornia crassipes, Myriophyllum spicatum, Phragmites australis, and Typha domingensis	Cd, Cu, Pb, and Zn	Water, sediment, and leaves, stems, and roots of the plants	Fawzy et al., 2012
Typha domingensis	Mg, Na, K, Cd, Cu, Fe, Pb, and Zn	Water, sediment, shoots, roots, and rhizomes of the plants	Eid et al., 2012
Eichhornia crassipes	Cu, Pb, and Hg	Shoots and roots of the plants	Padmapriya and Murugesan, 2012; Singh and Rai, 2016
Spirodela polyrrhiza	Cd, Fe, Pb, Cu, Zn, and Hg	Water and sediment	Loveson et al., 2013
Phragmites australis	Cd, Co, Cr, Cu, Mn, Ni, Pb, Zn, Sr, and V	Water, sediment, and leaves, stems, and roots of the plants	Kastratović et al., 2013
Myriophyllum verticillatum	Pb, Cu, Cd, and Zn	Whole plants	Ucer et al., 2013

(Continued)

TABLE 3.3 (CONTINUED)
Wetland Plants and Accumulated Metals (as Emerging Contaminants of Concern) in Phytoremediation

Scientific Names of Plants	Emerging Contaminant Names	Concentrated Area	References
Eichhornia crassipes	Fe, Al, Mn, Zn, Cu, Cr, As, Pb, Hg, Ni, and Cd	Leaf laminas, stems, petioles, and roots	Sasidharan et al., 2013; Singh and Rai, 2016
Ceratophyllum demersum, Myriophyllum spicatum, Eichhornia crassipes, Lemna gibba, Phragmites australis, and *Typha domingensis*	Cd, Co, Cu, Ni, Pb, and Zn	Water, sediments, and shoots and roots of the plants	Kamel, 2013
Juncus effuses and *Carex riparia*	Cd, Ni, and Zn	Urban storm water runoff, shoots and roots of the plants	Ladislas et al., 2013
Eichhornia crassipes, Hydrilla verticillata, Jussiaea repens, Lemna minor, Pistia stratiotes, and *Trapa natans*	Cu and Hg	Effluent water and the whole plants	Mishra et al., 2013; Singh and Rai, 2016
Phragmites australis	Al, Mn, Ba, Zn, Cu, Pb, Mo, Co, Cr, Cd, and Ni	Water, shoots and roots of the plants	Ahmed et al., 2014
Typha latifolia and *Phragmites australis*	As, B, Cd, Cr, Cu, Fe, Mn, Ni, Se, Sn, and Zn	Flowers, leaves, stems, rhizomes, and roots	Salem et al., 2014
T. domingensis, Ludwigia sp., *Paspalum vaginatum,* and *Paspalum stratiotes*	Cs, As, Hg	Root and shoot	Gbogbo and Otoo, 2015
Ipomoea aquatica Forsk	Pb	Root and shoot	Chanu and Gupta, 2016
Colocasia esculenta and *Scirpus articulates*	Cd, Cu, and Pb	Root and shoot	Khatu, 2016
Potamogeton–Ceratophyllum combination and *Phragmites australis*	Emerging contaminants, especially metals	Root and shoot	Ahmad et al., 2014, 2016
Lemna minor	Cd, Cu, Pb, and Ni	Dry mass	Bokhari et al., 2016

WHAT HAPPENS WITH EMERGING CONTAMINANT–LOADED AND METAL-SATURATED WETLAND PLANT BIOMASS?

Environmental contaminants are cleaned by hyperaccumulator plants through phytoremediation through loading them in aboveground and belowground biomass. Nevertheless, there exists an environmental concern after phytoremediation, attributed to the probability of leakage of these contaminants into the environment. Thus, concerns over the ultimate fate and safe disposal of these contaminant-loaded biomasses puzzle phytotechnologists. The invasive nature of certain wetland plants, like water hyacinth, also needs to be managed sustainably (Rai, 2008b, 2012, 2013a, 2015a–d, 2016a; Rai and Singh, 2015; Rai and Chutia, 2016). Henceforth, after being phytoremediated the emerging contaminants and NPs are not completely eliminated from the total environment. The existing conventional methods for the removal of hyperaccumulator plants, like ashing, incineration, composting, and landfill, tend to recycle these organic (diverse pesticides) or inorganic contaminants (metals) into the environment, leading to secondary pollution (soil and water pollution). In this context, one approach discussed is the *designer plant approach*, in which emerging contaminants are degraded within the plant biomass itself with microbial assistance. Thus, the resulting biomass, free from environmental contaminants, can be reused depending on its inherent characteristics. To this end, I suggest a couple of ways to sustainably manage the biomass of potent phytoremediation tools, that is, to phytosynthesize NPs, specifically metal NPs, and to couple their biomass for sustainable green renewable energy production. These strategies concomitantly boost the economic returns of phytotechnologies. Further, phytomining may be another approach as an extension to phytoremediation technology. Phytomining of precious metals like Au, using thiocyanate and thiosulfate, dates back almost 20 years to successful research investigations at Massey University, New Zealand. Further, Ag and Au uptakes were investigated by a cascade of research done at the University of Texas, El Paso, by G. Torresdey's group.

FUTURE PROSPECTS OF PHYTOREMEDIATION: GENETIC ENGINEERING AND MOLECULAR BIOLOGY

The application of genetic engineering to modify plants for the uptake, transport, and sequestration of metal stand other emerging contaminants may open up new avenues for enhancing the efficiency of phytoremediation (Eapen and D'Souza, 2005; Gomes et al., 2016) (see Table 3.4). It is a promising way toward the improvement of phytoremediation efficiency, thereby enhancing the metal tolerance and accumulation properties of plants (Moffat, 1999). Applications of gene manipulation to increase the phytoremediation pace are called *genoremediation* (Mani and Kumar, 2014; Gomes et al., 2016), with implications in molecular biology and tools in emerging contaminants such as homeostasis gene induction, gene responsiveness for biotic and abiotic stresses, biodegradative enzymatic genes, and risk-mitigating genes (Singh et al., 2011; Gomes et al., 2016).

Transgenic plants, which detoxify and accumulate metals like As, Cd, Hg, Pb, and Se, have been developed (Eapen and D'Souza, 2005). In recent years, several key

TABLE 3.4

Genetic Engineering Implications in Plant and Emerging Contaminant Removal: Bioenergy and Phytoremediation Interrelationship with Demonstration of Multiple Benefits of Global Phytotechnologies

S. No.	Genetically Manipulated Plant/Microbe	Plant under Expression of Inserted Gene	Emerging Contaminants	Gene/Protein Expressed	Significant Outcome of Genetic Engineering/Remarks (Benefits to Energy Sector and Environmental Implications)	Reference
1	*Staphylococcus aureus* (bacteria)	*Populus alba* and *Populus tremula* var. *glandulosa*	Hg	Mercury reductase	Five times better Hg phytoremediation in transgenic poplar than nontransgenics; potent biomass producing; couples bioenergy and phytoremediation	Choi et al., 2007; Gomes et al., 2016
2	*E. coli* (bacteria)	*Populus deltoides*	Hg	Mercuric ion reductase (*merA*) gene	Better Hg ion phytoremediation in concert with bioenergy, charcoal, pulp, and C sequestration	Che et al., 2003
3	*Iris lactea* var. *chinensis*	*Arabidopsis thaliana* (agro-crop)	Cu	Metallothionein	Enhanced Cu phytoremediation; reduced H_2O_2 production; increased root length in transgenic model plant	Gu et al., 2015; Gomes et al., 2016
4	*E. coli* (bacteria)	*Populus canescens*	Zn	g-Glutamylcysteine synthetase	Enhanced metal (Zn) phytoremediation in conjunction with bioenergy, charcoal, pulp, and C sequestration	Bittsanszkya et al., 2015
5	*E. coli* (bacteria and potent gene manipulation tool)	*Populus tremula*, *Populus alba*	Cd	g-Glutamylcysteine synthetase	Enhanced Cd phytoremediation; lower O_2 and H_2O_2 thiols; proline, GSH, and oxidized form of glutathione increased all toward stress reduction	He et al., 2015; Gomes et al., 2016

(Continued)

TABLE 3.4 (CONTINUED)

Genetic Engineering Implications in Plant and Emerging Contaminant Removal: Bioenergy and Phytoremediation Interrelationship with Demonstration of Multiple Benefits of Global Phytotechnologies

S. No.	Genetically Manipulated Plant/Microbe	Plant under Expression of Inserted Gene	Emerging Contaminants	Gene/Protein Expressed	Significant Outcome of Genetic Engineering/Remarks (Benefits to Energy Sector and Environmental Implications)	Reference
6	Arabidopsis thaliana (agro-crop)	Brassica juncea (agro-crop)	Se	ATP sulfurylase; cystathionine–g-synthase	Two times the escalated phytoremediation efficiency in above- and below-ground parts; Se caused growth reduction in both control and transgenic plants, but reduction was more pronounced in the control; transgenics showed two- to threefold higher Se volatilization rates	Pilon-Smits et al., 1999; Van Huysen et al., 2003; Gomes et al., 2016
7	Bacteria/microbes	Populus tremula × Populus alba	Diverse organic pollutant (trichloroethylene, vinyl chloride, benzene, CCl_4, and chloroform)	Transporter protein synthesis by microbial gene	Increased phytoremediation with concurrent relevance to bioenergy, related additives, and C sequestration	Doty et al., 2000
8	Saccharomyces cerevisiae (yeast)	Nicotiana tabacum	Cd and Zn	Metallothionein	Enhanced phytoremediation	Daghan et al., 2013; Gomes et al., 2016
9	Bacteria/microbes	Populus trichocarpa	Organic pesticides and herbicides	g-Glutamyl cysteine synthetase	Enhanced phytoremediation of chloroacetanilide herbicides/organics with resulting biomass for bioenergy, pulp, charcoal, and C sequestration	Gullner et al., 2001

(Continued)

TABLE 3.4 (CONTINUED)

Genetic Engineering Implications in Plant and Emerging Contaminant Removal: Bioenergy and Phytoremediation Interrelationship with Demonstration of Multiple Benefits of Global Phytotechnologies

S. No.	Genetically Manipulated Plant/Microbe	Plant under Expression of Inserted Gene	Emerging Contaminants	Gene/Protein Expressed	Significant Outcome of Genetic Engineering/Remarks (Benefits to Energy Sector and Environmental Implications)	Reference
10	*Escherichia coli*	*Brassica juncea*	Cd	g-Glutamylcysteine synthetase; GSH synthetase	Higher Cd tolerance as well as phytoremediation in shoots (threefold higher)	Zhu et al., 1999a,b; Gomes et al., 2016
11	Bacteria	*Hybrid aspen* (*Populus tremula* × *Populus tremuloides*)	TNT	Bacterial nitroreductase (pnrA)	Increased TNT phytoremediation coupled with bioenergy, charcoal, pulp, and C sequestration	Van Dillewijn et al., 2008
12	*Spinacia oleracea* (leafy agro-crop/ vegetable)	*Nicotiana tabacum*	Cd, Se, Ni, Pb, and Cu	Cysteine synthase	Enhanced tolerance toward Cd, Se, and Ni; however, not much impact in Pb and Cu tolerance and phytoremediation	Kawahsima et al., 2004
13	*Trametes versicolor*	*Populus seiboldii* × *Populus grandidentata* (hybrid poplar)	Escalated bisphenol A phytoremediation	Manganese peroxidase gene in poplar	Phytoremediation coupled with bioenergy from poplar biomass and C sequestration	Iimura et al., 2007

steps have been identified at the molecular level, enabling us to initiate transgenic approaches to engineering the transition metal content of plants (Clemens et al., 2002). Complex interactions of transport and chelating activities control the rates of metal uptake and storage (Rai, 2009a). Metal chelator; metal transporter; metallo-thionein (MT), such as glutathione (GSH); and phytochelatin (PC) (Rai, 2009a) genes have been incorporated into plants for improved metal uptake and sequestration (Eapen and D'Souza, 2005). PCs are the family of peptides with the general structure (γ-Glu-Cys) n-Gly, where n equals 2–11 (Rauser, 1990; Cobbett, 2000); they are rapidly synthesized in response to toxic levels of heavy metals (Zenk, 1996; Cobbett, 1999; Rai, 2009a). PCs are enzymatically synthesized from GSH by phyto-chelatin synthase (PCS) (EC 2.3.2.15) (Cobbett, 2000; Rai, 2009a) and bind heavy metals such as Ag, Ar, Cd, or Cu (Maitani et al., 1996; Schmoger et al., 2000; Rai, 2009a). PCS genes were cloned from *Arabidopsis thaliana, Caenorabditi elegans, Schizosaccharomyces pombe,* and *Triticum aestivum* (Clemens et al., 1999, 2002; Ha et al., 1999; Vatamaniuk et al., 1999; Rai, 2009a; Singh and Rai, 2016).

Plants such as *Populus angustifolia, Nicotiana tabacum,* and *Silene cucubalis* have been genetically engineered to overexpress glutamylcysteine synthetase, which enhances heavy metal accumulation compared with similar wild plants (Fulekar et al., 2009). Nevertheless, there exists scanty research on genetic engineering and molecular biology in the context of wetland plants or macrophytes.

To this end, *Typha latifolia* was used for genetic manipulation using a standard-ized *Agrobacterium*-mediated model transformation system to achieve the long-term objective of introducing candidate genes for phytoremediation (Nandakumar et al., 2005; Rai, 2009a). Further, a cDNA encoding a type 2 MT was isolated from *Azolla filiculoides,* termed AzMT2, accession no. AF482470, and AzMT2 RNA expression was boosted by the addition of Cd, Cu, Zn, and Ni to the growth medium (Fumbarov et al., 2005; Rai, 2009a). Moreover, in the context of wetland plants, molecular tools like x-ray absorption spectroscopy are now being applied for *Eichhornia crassipes* and *Azolla filiculoides* to investigate metal interactions within biosystems and to study important information on the coordination chemistry of metals and toxic ele-ment interactions with phytoremediation systems (Gardea-Torresdey et al., 2005; Rai, 2009a).

The use of phytoremediation technology using wetland macrophyte biore-sources should be encouraged in order to replace costly, ineffective, and conven-tional technologies. It is well suited for a developing country like India, where there is huge water pollution, especially in the wetlands. The use of invasive plant species like *Eichhornia crassipes* for this technology is also an important concept for utilizing freely available plant resources (Rai and Singh, 2016). The remedia-tion of hazardous heavy metals and other toxic contaminants from the wetlands through wetland bioresources will benefit the associated living beings in providing them a toxic-free, healthy environment. A concerted multidisciplinary approach integrating wetland ecology, pollution science, environmental engineering, genetic engineering, and sustainability issues with macrophyte bioresources may assist in green phytotechnological innovation. Microbial efficiency inextricably linked with wetland plants can be enhanced for escalating phytotechnologies and has a potentially bright future.

REFERENCES

Ahmad S et al. 2016. Heavy metal accumulation in the leaves of *Potamogeton natans* and *Ceratophyllum demersum* in a Himalayan RAMSAR site: Management implications. *Wetlands Ecol Manag* 24:469–475.

Ahmad SS, Reshi ZA, Shah MA, Rashid I, Ara R, Andrabi SMA. 2014. Phytoremediation potential of *Phragmites australis* in Hokersar wetland—A Ramsar site of Kashmir Himalaya. *Int J Phytoremediation* 16(12):1183–1191.

Alam MGM, Snow ET, Tanaka A. 2003. Arsenic and heavy metal contamination of vegetables grown in Samta Village, Bangladesh. *Sci Total Environ* 308:83–96.

Alhashemi AH, Karbassi AR, Kiabi BH, Monavari SM, Nabavi MB. 2011. Accumulation and bioaccessibility of trace elements in wetland sediments. *Afr J Biotechnol* 10(9):1625–1638.

Ali D, Nagpure NS, Kumar S, Kumar R, Kushwaha B, Lakra WS. 2009. Assessment of genotoxic and mutagenic effects of chlorpyrifos in freshwater fish *Channa punctatus* (Bloch) using micronucleaus assay and alkaline single-cell gel electrophoresis. *Food Chem Toxicol* 47(3):650–656.

Al-Masri MS, Aba A, Khalil H, Al-Hares Z. 2002. Sedimentation rates and pollution history of a dried lake: Al-Qteibeh Lake. *Sci Total Environ* 293(1–3):177–189.

Arroyo P, Blanco I, Cortijo R, Calabuig EL, Ansola G. 2013. Twelve-year performance of a constructed wetland for municipal wastewater treatment: Water quality improvement, metal distribution in wastewater, sediments and vegetation. *Water Air Soil Pollut* 224:1762.

Badejo A, Sridhar MKC, Coker AO, Ndambuki JM, Kupolati WK. 2015. Phytoremediation of water using *Phragmites karka* and *Veteveria nigritana* in constructed wetland. *Int J Phytoremediation* 17(9):847–852.

Bako SP, Daudu P. 2007. Trace metal contents of the emergent macrophytes *Polygonum* sp. and *Ludwigia* sp. in relation to the sediments of two freshwater lake ecosystems in the Nigerian savanna. *J Fish Aquat Sci* 2(1):63–70.

Baldantoni D, Alfani A, Di Tommasi P, Bartoli G, De Santo A. 2004. Assessment of macro and microelement accumulation capability of two aquatic plants. *Environ Pollut* 130:149–156.

Baldantoni D, Maisto G, Bartoli G, Alfani A. 2005. Analyses of three native aquatic plant species to assess spatial gradients of lake trace element contamination. *Aquat Bot* 83:48–60.

Balsberg-Pahlsson AM. 1989. Toxicity of heavy metal (Zn, Cu, Cd, Pb) to vascular plants: A literature review. *Water Air Soil Pollut* 47:287–319.

Bassi N, Kumar MD, Sharma A, Pardha-Saradhi P. 2014. Status of wetlands in India: A review of extent, ecosystem benefits, threats and management strategies. *J Hydrol Regional Stud* 2:1–19.

Begam A, HariKrishna S. 2010. Bioaccumulation of trace metals by aquatic plants. *Int J Chemtech Res* 2(1):250–254.

Berman E. 1980. *Toxic Metals and Their Analysis*. London: Heyden.

Bickham JW, Sandhu S, Hebert PDN, Chikhi L, Athwal R. 2000. Effects of chemical contaminations on genetic diversity in natural populations: Implication for biomonitoring and ecotoxicology. *Mutat Res* 463:33–51.

Bittsanszkya A et al. 2005. Ability of transgenic poplars with elevated glutathione content to tolerate zinc(2+) stress. *Environ Int* 31:251–254.

Bokhari S, Ahmad I, Mahmood-Ul-Hassan M, Mohammad A. 2016. Phytoremediation potential of *Lemna minor* L. for heavy metals. *Int J Phytoremediation* 18(1):25–32.

Boudet LC, Escalante A, von Haeften G, Moreno V, Gerpe M. 2011. Assessment of heavy metals accumulation in two aquatic macrophytes: A field study. *J Braz Soc Ecotoxicol* 6(1):57–64.

Branković S, Pavlović-Muratspahić D, Topuzović M, Glišić R, Banković D, Stanković M. 2011. Environmental study of some metals on several aquatic macrophytes. *Afr J Biotechnol* 10(56):11956–11965.

Bystrom O, Andersson H, Gren I. 2000. Economics criteria for using wetlands as nitrogen sinks under uncertainty. *Ecol Econ* 35(1):35–45.

Cairns J, Smith EP, Orvos D. 1988. The problem of validating simulation of hazardous exposure in natural systems. In *Society for Computer Simulation International 1988 Summer Computer Conference Proceedings*, San Diego, CA, pp. 448–454.

Chandra AS, Rath P, Chandra PU, Kumar PP, Brahma S. 2013. Application of sequential leaching, risk indices and multivariate statistics to evaluate heavy metal contamination of estuarine sediments: Dhamara Estuary, east coast of India. *Environ Monit Assess* 185:6719–6737.

Chanu LB, Gupta A. 2016. Phytoremediation of lead using *Ipomoea aquatica* Forsk. in hydroponic solution. *Chemosphere* 156:407–411.

Chatterjee S, Chetia M, Singh L, Chattopadyay B, Datta S, Mukhopadhyay SK. 2011. A study on the phytoaccumulation of waste elements in wetland plants of a Ramsar site in India. *Environ Monit Assess* 178:361–371.

Che D et al. 2003. Expression of mercuric ion reductase in Eastern cottonwood (*Populus deltoides*) confers mercuric ion reduction and resistance. *J Plant Biotechnol* 1:311–319.

Choi YI et al. 2007. Mercury-tolerant transgenic poplars expressing two bacterial mercury-metabolizing genes. *J Plant Biol* 50:658.

Clemens S, Kim EJ, Neumann D, Schroeder JI. 1999. Tolerance to toxic metals by a gene family of phytochelatin synthases from plants and yeast. *EMBO J* 18:3325–3333.

Clemens S, Palmgren MG, Krämer U. 2002. A long way ahead: Understanding and engineering plant metal accumulation. *Trends Plant Sci* 7(7):309–315.

Cobbett CS. 1999. A family of phytochelatin synthase genes from plant, fungal and animal species. *Trends Plant Sci* 4:335–337.

Cobbett CS. 2000. Phytochelatins and their roles in heavy metal detoxification. *Plant Physiol* 123:825–832.

Concas A, Ardau C, Cristini A, Zuddas P, Cao G. 2006. Mobility of heavy metals from tailings to stream waters in mining activity contaminated site. *Chemosphere* 63:244–253.

Costanza R et al. 1997. The value of the world's ecosystem services and natural capital. *Nature* 387(6630):253–260.

Cowardin LM, Carter V, Golet FC, LaRoe ET. 1979. *Classification of Wetlands and Basis.* Washington, DC: Wildlife Service.

Daghan H et al. 2013. Transformation of tobacco with ScMTII gene-enhanced cadmium and zinc accumulation. *Clean Soil Air Water* 41:503–509.

Das S, Goswami S, Talukdar AD. 2014. A study on cadmium phytoremediation potential of water lettuce, *Pistia stratiotes* L. *Bull Environ Contam Toxicol* 92:169–174.

Devi BN, Sharma BM. 2002. Life form analysis of the macrophytes of the Loktak Lake, Manipur, India. *Indian J Environ Ecoplan* 6(3):451–458.

Doherty JM, Miller JF, Prellwitz SG, Thompson AM, Loheide SP II, Zedler JB. 2014. Hydrologic regimes revealed bundles and tradeoffs among six wetland services. *Ecosystems* 17:1026–1039.

Doty SL et al. 2000. Enhanced metabolism of halogenated hydrocarbons in transgenic plants containing mammalian cytochrome P450 2E1. *Proc Natl Acad Sci USA* 97:6287–6291.

Dsikowitzky L, Nordhaus L, Jennerjahn TC, Khrycheva P, Sivatharshan Y, Yuwono E, Schwarzbauer J. 2011. Anthropogenic organic contaminants in water, sediments and benthic organisms of the mangrove-fringed Segara Anakan Lagoon, Java, Indonesia. *Marine Pollut Bull* 62:851–862.

Eapen S, D'Souza SF. 2005. Prospect of genetic engineering of plants for phytoremediation of toxic metals. *Biotechnol Adv* 23(2):97–114.

Eccles, H. 1999. Treatment of metal-contaminated wastes: Why select a biological process? *Trends Biotechnol* 17:462–465.

Eid EM, Shalhout KH, El-Sheikh M, Asaeda T. 2012. Seasonal courses of nutrients and heavy metals in water, sediment and above- and below-ground *Typha domingensis* biomass in Lake Burullus (Egypt): Perspectives for phytoremediation. *Flora* 207:783–794.

EPA (Environmental Protection Agency). 2004. Environmental protection of wetlands. Washington, DC: EPA.

Fairbrother A, Wenstel R, Sappington S, Wood W. 2007. Framework for metals risk assessment. *Ecotoxicol Environ Saf* 68:145–227.

Farraji H, Zaman NQ, Tajuddin RM, Faraji H. 2016 Advantages and disadvantages of phytoremediation: A concise review. *Int J Environ Technol Sci* 2:69–75.

Fatima M, Usmani N, Hossain MM, Siddiqui MF, Zafeer MF, Firdaus F, Ahmed S. 2014. Assessment of genotoxic induction and deterioration of fish quality in commercial species due to heavy-metal exposure in an urban reservoir. *Arch Environ Contam Toxicol* 67:203–213.

Fawzy MA, Badr NES, Khatib AE, Kassem AAE. 2012. Heavy metal biomonitoring and phytoremediation potentialities of aquatic macrophytes in River Nile. *Environ Monit Assess* 184:1753–1771.

Feng L, Wen YM, Zhu PT. 2008. Bioavailability and toxicity of heavy metals in a heavily polluted river in PRD, China. *Contam Toxicol* 81:90–94.

Finkelman RB, Gross PMK. 1999. The types of data needed for assessing the environmental and human health impacts of coal. *Int J Coal Geol* 40:91–101.

Forni C, Tommasi F. 2015. Duckweed: A tool for ecotoxicology and a candidate for phytoremediation. *Curr Biotechnol* 4:000–000.

Fumbarov TS, Goldsbrough PB, Adam Z, Tel-Or E. 2005. Characterization and expression of a metallothionein gene in the aquatic fern *Azolla filiculoides* under heavy metal stress. *Planta* 223:69–76.

Gardea-Torresdey JL, Peralta-Videa JR, Rosa GDL, Parson JG. 2005. Phytoremediation of heavy metals and study of the metal coordination by x-ray spectroscopy. *Coord Chem Rev* 17–18:1797–1810.

Ghermandi A, van der Bergh JCJM, Brander LM, Nunes PALD. 2008. The economics value of wetland conservation and creation: A meta-analysis. Working Paper 79. Milan: Fondazione Eni Enrici Mattei.

Ginneken LV, Meers E, Guisson R, Ruttens A, Elst K, Tack FMG, Vangronsveld J, Dejonghe W. 2007. Phytoremediation for heavy meta-contaminated soils combined with bioenergy production. *J Environ Eng Landscape Manag* 15(4):227–236.

Glover-Kerkvilet J. 1995. Environmental assault on immunity. *Environ Health Perspect* 103:236–237.

Gomes HI. 2012. Phytoremediation for bioenergy: Challenges and opportunities. *Environ Technol Rev* 1(1):59–66.

Gomes MAC et al. 2016. Metal phytoremediation: General strategies, genetically modified plants and applications in metal nanoparticle contamination. *Ecotoxicol Environ Saf* 134:133–147.

Gu CS et al. 2015. The heterologous expression of the *Iris lacteal* var. *chinensis* type 2 metallothionein IlMT2b gene enhances copper tolerance in *Arabidopsis thaliana. Bull Environ Contam Toxicol* 94:247–253.

Guilizzoni P. 1991. The role of heavy metals and toxic materials in the physiological ecology of submersed macrophytes. *Aquat Bot* 41:87–109.

Gullner G et al. 2001. Enhanced tolerance of transgenic poplar plants overexpressing gamma-glutamylcysteine synthetase towards chloroacetanilide herbicides. *J Exp Bot* 52:971–979.

Gurcu B, Yildiz S, Koca YBG, Koca S. 2010. Investigation of histopathological and cytogenetic effects of heavy metals pollution on *Cyprinus carpino* (Linneaus, 1758) in the Gölmar-mara Lake, Turkey. *J Anim Vet Adv* 9:798–808.

Ha SB, Smith AP, Howden R, Dietrich WM, Bugg S, O'Connell MJ, Goldsbrough PB, Cobbett CS. 1999. Phytochelatin synthase genes from *Arabidopsis* and the yeast *Schizosaccharomyces pombe*. *Plant Cell* 11:1153–1164.

Haddam N, Samira S, Dumon X, Taleb A, Lison D, Haufroid V, Bernard A. 2011. Confounders in the assessment of the renal effects associated with low-level urinary cadmium: An analysis in industrial workers. *Environ Health* 1037. doi: 10.1186/1476-069X-10-37.

Hanninen H, Lindstrom H. 1979. Behavioral test battery for toxic psychological studies used at the Institute of Occupational Health at Helsinki. Helsinki: Institute of Occupational Health.

Hänsch R, Mendel RR. 2009. Physiological functions of mineral micronutrients (Cu, Zn, Mn, Fe, Ni, Mo, B, Cl). *Curr Opin Plant Biol* 12:259–266.

Harikumar PS, Nasir UP, Mujeebu Rahman MP. 2009. Distribution of heavy metals in the core sediments of a tropical wetland system. *Int J Environ Sci Technol* 6(2):225–232.

He J et al. 2015. Overexpression of bacterial g-glutamylcysteine synthetase mediates changes in cadmium influx, allocation and detoxification in poplar. *New Phytol* 205:240–254.

Ho KC, Chow YL, Yau JTS. 2003. Chemical and microbiological qualities of the East River (Dongjiang) water, with particular reference to drinking water supply in Hong Kong. *Chemosphere* 52:1441–1450.

Iimura Y et al. 2007. Hybrid aspen with a transgene for fungal manganese peroxidase is a potential contributor to phytoremediation of the environment contaminated with bisphenol A. *J Wood Sci* 53:541–544.

Jain CK, Singhal DC, Sharma MK. 2007. Estimating nutrient loadings using chemical mass balance approach. *Environ Monit Assess* 134(1–3):385–396.

Jarup L. 2003. Hazards of heavy metal contamination. *Br Med Bull* 68:167–182.

Jiao W, Ouyang W, Hao FH, Wang FL, Liu B. 2014. Long-term cultivation impact on the heavy metal behaviour in a reclaimed wetland, Northeast China. *Environ Pollut* 147:311–323.

Kamel KA. 2013. Phytoremediation potentiality of aquatic macrophytes in heavy metal contaminated water of E-Temsah Lake, Ismailia, Egypt. *Middle East J Sci Res* 14(12):1555–1568.

Kastratović V, Krivokapić S, Đurović D, Blagojević N. 2013. Seasonal changes in metal accumulation and distribution in the organs of *Phragmites australis* (common reed) from Lake Skadar, Montenegro. *J Serb Chem Soc* 78(8):1241–1258.

Kawashima CG et al. 2004. Heavy metal tolerance of transgenic tobacco plants overexpressing cysteine synthase. *Biotechnol Lett* 26:153–157.

Khatun A. 2016. Evaluation of metal contamination and phytoremediation potential of aquatic macrophytes of East Kolkata Wetlands, India. *Environ Health Toxicol* 31:7.

Knasmuller S, Gottmann E, Steinkellner H, Fomin A, Pickl C, Paschke A, God R, Kundli M. 1998. Detection of genotoxic effects of heavy metal contaminated soils with plant bioassay. *Mutat Res* 420:37–48.

Kudo A, Miyahara S. 1991. A case history: Minamata mercury pollution in Japan—From loss of human lives to decontamination. *Water Sci Technol* 23:283.

Kumar A, Pastore P. 2007. Lead and cadmium in soft plastic toys. *Curr Sci India* 93:818–822.

Kumar JIN, Soni R, Kumar RN, Bhatt I. 2008. Macrophytes in phytoremediation of heavy metal contaminated water and sediments in Periyej Community Reserve, Gujarat, India. *Turk J Fish Aquat Sci* 8:193–200.

Ladislas S, Gérente C, Chazarenc F, Brisson J, Andrès Y. 2013. Performances of two macrophytes species in floating treatment wetlands for cadmium, nickel, and zinc removal from urban stormwater runoff. *Water Air Soil Pollut* 224:1408.

Lauwerys RR. 1979. *Health Effects of Cadmium*, 43–64. Oxford: Pergamon Press.

LDA (Loktak Development Authority). 2011. Annual administrative report 2010–11. Manipur, India: LDA.

Lee J, Rai PK, Jeon YJ, Kim KH, Kwon EE. 2017. The role of algae and cyanobacteria in the production and release of odorants in water. *Environ Pollut* 227:252–262.

Li YL, Liu YG, Liu JL, Zeng GM, Li X. 2008. Effects of EDTA on lead uptake by *Typha orientalis* Presl: A new lead-accumulating species in southern China. *Bull Environ Contam Toxicol* 81:36–41.

Lin HJ, Sunge T, Cheng CY, Guo HR. 2013. Arsenic levels in drinking water and mortality of liver cancer in Taiwan. *J Hazard Mater* 262:1132–1138.

Liu JG, Diamond J. 2005. China's environment in a globalizing world. *Nature* 435:1179–1186.

Liu JL, Li YL, Zhang B, Cao JL, Cao ZG, Domagalski J. 2009. Ecological risk of heavy metals in sediments of the Luan River source water. *Ecotoxicology* 18:748–758.

Liu Y, Chen M, Jiang L, Song L. 2014. New insight into molecular interaction of heavy metal pollutant-cadmium (II) with human serum albumin. *Environ Sci Pollut Res* 21:6994–7005.

Loveson A, Sivalingam R, Syamkumar R. 2013. Aquatic macrophyte *Spirodela polyrrhiza* as a phytoremediation tool in polluted wetland water from Eloor, Ernakulam District, Kerala. *J Environ Anal Toxicol* 3(5):1–7.

Maitani T, Kubota H, Sato K, Yamada T. 1996. The composition of metals bound to class III metallothionein (phytochelatin and its desglycyl peptide) induced by various metals in root cultures of *Rubia tinctorum*. *Plant Physiol* 110:1145–1150.

Malar S, Sahi SV, Favas PJC, Venkatachalam P. 2015. Mercury heavy-metal-induced physiochemical changes and genotoxic alterations in water hyacinths [*Eichhornia crassipes* (Mart.)]. *Environ Sci Pollut Res* 22:4597–4608.

Mander Ü, Mitsch WJ. 2009. Pollution control by wetlands. *Ecol Eng* 35(2):153–158.

Mani D, Kumar C. 2014. Biotechnological advances in bioremediation of heavy metals contaminated ecosystems: An overview with special reference to phytoremediation. *Int J Environ Sci Technol* 11(3):843–872.

Mayon N et al. 2006. Multiscale approach of fish responses to different types of environmental contaminations: A case study. *Sci Total Environ* 367:715–773.

McLaughlin MJ, Parker DR, Clark JM. 1999. Metals and micronutrients—Food safety issues. *Field Crops Res* 60:143–163.

Mdegela RH, Braathen M, Pereka AE, Mosha RD, Sandvik M, Skaare JU. 2009. Heavy metals and organochlorine residues in water, sediments, and fish in aquatic ecosystems in urban and peri-urban areas in Tanzania. *Water Air Soil Pollut* 203:369–379.

Mengel K, Kirkby EA, Kosegarten H, Appel T. 2001. *Principles of Plant Nutrition*. Dordrecht, the Netherlands: Kluwer.

Miller JR, Hudson-Edwards KA, Lechler PJ, Preston D, Macklin MG. 2004. Heavy metal contamination of water, soil and produce within riverine communities of the Rio Pilcomayo basin, Bolivia. *Sci Total Environ* 320:189–209.

Mishra S, Mohanty M, Pradhan C, Patra HK, Das R, Sahoo S. 2013. Physico-chemical assessment of paper mill effluent and its heavy metal remediation using aquatic macrophytes—A case study at JK Paper Mill, Rayagada, India. *Environ Monit Assess* 185:4347–4359.

Moffat AS. (1999). Engineering plants to cope up with metals. *Science* 285:369–370.

Montuori P, Lama P, Aurino S, Naviglio D, Triassi M. 2013. Metals loads into the Mediterranean Sea: Estimate of Sarno River inputs and ecological risk. *Ecotoxicology* 22:295–307.

Moreno-Mateos D, Power ME, Comin FA, Yockteng R. 2012. Structural and functional loss in restored wetland ecosystems. *PLoS Biol* 10:e1001247.

Nandakumar R, Chen L, Rogers SMD. 2005. *Agrobacterium*-mediated transformation of the wetland monocot *Typha latifolia* L. (broadleaf cattail). *Plant Cell Rep* 23:744–750.

Nies DH. 1999. Microbial heavy-metal resistance. *Appl Microbiol Biotechnol* 51:730–750.

Noegrohati S. 2005. Sorption–desorption characteristics of heavy metals and their availability from the sediment of Segara anakan estuary. *Indones J Chem* 5(3):236–244.

Nordberg GF. 1996. Current issues in low-dose cadmium toxicology: Nephrotoxicity and carcinogenicity. *Environ Sci* 4(3):133–147.

Nriagu JO. 1979. Global inventory of natural and anthropogenic emission of trace metals to the atmosphere. *Nature* 279:409–411.

Nriagu JO. 1996. A history of global metal pollution. *Science* 272:273–274.

Nursita AI, Singh B, Lees E. 2009. Cadmium bioaccumulation in *Proisotoma minuta* in relation to bioavailability in soils. *Ecotoxicol Environ Saf* 72:1767–1773.

Oberholster PJ, McMillan P, Durgapersad K, Botha AM, de Klerk AR. 2014. The development of a wetland classification and risk assessment index (WCRAI) for non-wetland specialists for the management of natural freshwater wetlands ecosystems. *Water Air Soil Pollut* 225:1833.

Padmapriya G, Murugesan AG. 2012. Phytoremediation of various heavy metals (Cu, Pb and Hg) from aqueous solution using water hyacinth and its toxicity on plants. *Int J Environ Biol* 97–103.

Pandey SK, Tripathi BD, Prajapati SK, Mishra VK, Upadhyay AR, Rai PK, Sharma AP. 2005. Magnetic properties of vehicle derived particulates and amelioration by *Ficus infectoria*: A keystone species. *Ambio* 34(8):645–646.

PCD (Pollution Control Department). 2000. Groundwater standards for drinking proposes. Manual inspection of contaminated groundwater standards for drinking proposes. Manual inspection of contaminated groundwater from waste disposal facilities. Ministry of Natural Resources and Environment.

Pilon-Smits E, Pilon M. 2002. Phytoremediation of metals using transgenic plants. *Crit Rev Plant Sci* 21(5):439–456.

Pilon-Smits EAH et al. 1999. Overexpression of ATP sulfurylase in Indian mustard leads to increased selenate uptake, reduction, and tolerance. *Plant Physiol* 119:123–132.

Pip E, Stepaniuk J. 1992. Cadmium, copper and lead in sediments and aquatic macrophytes in the Lower Nelson River System, Manitoba, Canada. I. Interspecific differences and macrophyte–sediment relations. *Arch Hydrobiol* 124:337–355.

Prasad SN, Ramachandra TV, Ahalya N, Sengupta T, Kumar A, Tiwari AK, Vijayan VS, Vijayan L. 2002. Conservation of wetlands of India—A review. *Trop Ecol* 43(1):173–186.

Rai PK. 2007a. Phytoremediation of Pb and Ni from industrial effluents using *Lemna minor*: An eco-sustainable approach. *Bull BioSci* 5(1):67–73.

Rai PK. 2007b. Wastewater management through biomass of *Azolla pinnata*: An ecosustainable approach. *Ambio* 36(5):426–428.

Rai PK. 2008a. Phytoremediation of Hg and Cd from industrial effluents using an aquatic free floating macrophyte *Azolla pinnata*. *Int J Phytoremediation* 10(5):430–439.

Rai PK. 2008b. Heavy-metal pollution in aquatic ecosystems and its phytoremediation using wetland plants: An ecosustainable approach. *Int J Phytoremediation* 10(2):133–160.

Rai PK. 2008c. Mercury pollution from chlor-alkali industry in a tropical lake and its biomagnification in aquatic biota: Link between chemical pollution, biomarkers and human health concern. *Human Ecol Risk Assess Int J* 14:1318–1329.

Rai PK. 2009a. Heavy metal phytoremediation from aquatic ecosystems with special reference to macrophytes. *Crit Rev Environ Sci Technol* 39(9):697–753.

Rai PK. 2010a. Microcosm investigation on phytoremediation of Cr using *Azolla pinnata*. *Int J Phytoremediation* 12:96–104.

Rai PK. 2010b. Phytoremediation of heavy metals in a tropical impoundment of industrial region. *Environ Monit Assess* 165:529–537.

Rai PK. 2010c. Seasonal monitoring of heavy metals and physico-chemical characteristics in a lentic ecosystem of sub-tropical industrial region, India. *Environ Monit Assess* 165:407–433.

Rai PK. 2010d. Heavy metal pollution in lentic ecosystem of sub-tropical industrial region and its phytoremediation. *Int J Phytoremediation* 12(3):226–242.

Rai PK. 2011. *Heavy Metal Pollution and Its Phytoremediation through Wetland Plants*. New York: Nova Science Publisher.

Rai PK. 2012. Assessment of multifaceted environmental issues and model development of an Indo-Burma hot spot region. *Environ Monit Assess* 184:113–131.

Rai PK. 2013. Environmental magnetic studies of particulates with special reference to bio-magnetic monitoring using roadside plant leaves. *Atmos Environ* 72:113–129.

Rai PK. 2013a. *Plant Invasion Ecology: Impacts and Sustainable Management*. New York: Nova Science Publisher.

Rai PK. 2013b. Environmental magnetic studies of particulates with special reference to bio-magnetic monitoring using roadside plant leaves. *Atmos Environ* 72:113–129.

Rai PK. 2015a. Paradigm of plant invasion: Multifaceted review on sustainable management. *Environ Monit Assess* 187:759–785.

Rai PK. 2015b. What makes the plant invasion possible? Paradigm of invasion mechanisms, theories and attributes. *Environ Skeptics Critics* 4(2):36–66.

Rai PK. 2015c. Concept of plant invasion ecology as prime factor for biodiversity crisis: Introductory review. *Int Res J Environ Sci* 4(5):85–90.

Rai PK. 2015d. *Environmental Issues and Sustainable Development of North East India*. Saarbrücken, Germany: Lambert Academic Publisher.

Rai PK. 2015e. Multifaceted health impacts of particulate matter (PM) and its management: An overview. *Environ Skeptics Critics* 4(1):1–26.

Rai PK. 2016a. *Biomagnetic Monitoring through Roadside Plants of an Indo-Burma Hot Spot Region*. Oxford: Elsevier.

Rai PK. 2016b. Biodiversity of roadside plants and their response to air pollution in an Indo-Burma hotspot region: Implications for urban ecosystem restoration. *J Asia Pac Biodivers* 9:47–55.

Rai PK. 2016c. Impacts of particulate matter pollution on plants: Implications for environmental biomonitoring. *Ecotoxicol Environ Saf* 129:120–136.

Rai PK, Chutia B. 2016. Biomagnetic monitoring through *Lantana* leaves in an Indo-Burma hot spot region. *Environ Skeptics Critics* 5(1):1–11.

Rai PK, Chutia BM, Patil SK. 2014. Monitoring of spatial variations of particulate matter (PM) pollution through bio-magnetic aspects of roadside plant leaves in an Indo-Burma hot spot region. *Urban For Urban Green* 13:761–770.

Rai PK, Mishra, A, Tripathi BD. 2010. Heavy metals and microbial pollution of river Ganga: A case study on water quality at Varanasi. *Aquat Ecosyst Health Manag* 13(4):352–361.

Rai PK, Panda LS. 2014. Dust capturing potential and air pollution tolerance index (APTI) of some roadside tree vegetation in Aizawl, Mizoram, India: An Indo-Burma hot spot region. *Air Qual Atmos Health* 7(1):93–101.

Rai PK, Panda LS, Chutia BM, Singh MM. 2013. Comparative assessment of air pollution tolerance index (APTI) in the industrial (Rourkela) and non-industrial area (Aizawl) of India: An eco-management approach. *Afr J Environ Sci Technol* 7(10):944–948.

Rai PK, Singh MM. 2015. *Lantana camara* invasion in urban forests of an Indo-Burma hotspot region and its ecosustainable management implication through biomonitoring of particulate matter. *J Asia Pac Biodivers* 8:375–381.

Rai PK, Sharma AP, Tripathi BD. 2007. Urban environment status in Singrauli Industrial region and its eco-sustainable management: A case study on heavy metal pollution. In *Urban Planing and Environment, Strategies and Challenges*, ed. Vyas L, 213–217. McMillan Advanced Research Series.

Rai PK, Singh M. 2016. *Eichhornia crassipes* as a potential phytoremediation agent and an important bioresource for Asia Pacific region. *Environ Skeptics Critics* 5(1):12–19.

Rai PK, Tripathi BD. 2008. Heavy metals in industrial wastewater, soil and vegetables in Lohta village, India. *Toxicol Environ Chem* 90(2):247–257.

Rai PK, Tripathi BD. 2009. Comparative assessment of *Azolla pinnata* and *Vallisneria spiralis* in Hg removal from G.B. Pant Sagar of Singrauli Industrial region, India. *Environ Monitor Assess* 148:75–84.

Rai PK et al. 2018. A critical review of ferrate(VI)-based remediation of soil and groundwater. *J Environ Res* 160:420–448.

Rai SC, Raleng A. 2011. Ecological studies of wetland ecosystem in Manipur valley from management perspectives. In *Ecosystems Biodiversity*, ed. O Grillo, G Venora, 233–248. Rijeka, Croatia: InTech.

Rai UN, Tripathi RD, Vajpayee P, Vidyanath J, Ali MB. 2002. Bioaccumulation of toxic metals (Cr, Cd, Pb and Cu) by seeds of *Euryle ferox* Salisb (Makhana). *Chemosphere* 46:267–272.

Rattan RK, Datta SP, Chhonkar PK, Suribabu K, Singh AK. 2005. Long-term impact of irrigation with sewage effluents on heavy metal content in soils, crops and groundwater—A case study. *Agric Ecosyst Environ* 109:310–322.

Rauser WE. 1990. Phytochelatins. *Annu Rev Biochem* 59:61–86.

Roy DR. 1992. Case study of Loktak Lake of Manipur. In *Wetlands of India*, ed. KJS Chatrath. New Delhi: Ashish Publishing House.

Salem ZB, Laffray X, Ashoour A, Ayadi H, Aleya L. 2014. Metal accumulation and distribution in the organs of reeds and cattails in a constructed treatment wetland (Etueffont, France). *Ecol Eng* 64:1–17.

Sanjit L, Bhatt D, Sharma RK. 2005. Habitat heterogeneity of the Loktak Lake, Manipur. *Curr Sci* 88(7):1028.

Sasidharan NK, Azim T, Devi DA, Mathew S. 2013. Water hyacinth for heavy metal scavenging and utilization as organic manure. *Indian J Weed Sci* 45(3):204–209.

Sasmaz A, Obek E, Hasar H. 2008. The accumulation of heavy metals in *Typha latifolia* L. grown in a stream carrying secondary effluent. *Ecol Eng* 33:278–284.

Schmoger ME, Oven M, Grill E. 2000. Detoxification of arsenic by phytochelatins in plants. *Plant Physiol* 122:793–801.

Sharma DC. 2003. Concern over mercury pollution in India. *Lancet* 362:1030.

Sheoran AS, Sheoran V. 2006. Heavy metal removal mechanism of acid mine drainage in wetlands: A critical review. *Min Eng* 19:105–116.

Shukla VK, Prakash A, Tripathi BD, Reddy DCS. 1998. Biliary heavy metal concentration in carcinoma of the gall bladder: Case-control study. *BMJ* 317:1288–1289.

Singh A, Prasad SM. 2015. Remediation of heavy metal contaminated ecosystem: An overview on technology advancement. *Int J Environ Sci Technol* 12:353–266.

Singh JS et al. 2011. Genetically engineered bacteria: An emerging tool for environmental remediation and future research perspectives. *Gene* 480:1–9.

Singh MM, Rai PK. 2016. Microcosm investigation of Fe (iron) removal using macrophytes of Ramsar lake: A phytoremediation approach. *Int J Phytoremediation* 18(12):1231–1236.

Singh MR, Gupta A, Beeteswari KH. 2010. Physicochemical properties of water samples from Manipur River System, India. *J Appl Sci Environ Manag* 14(4):85–89.

Singh NKS, Sudarshan M, Chakraborty A, Devi ChB, Singh ThB, Singh NR. 2014. Biomonitoring of fresh water of Loktak Lake, India. *Eur J Sustain Dev* 3(1):179–188.

Sood A, Uniyal PP, Prasanna, R, Ahluwalia AS. 2012. Phytoremediation potential of aquatic macrophyte. *Azolla. Ambio* 41:122–137.

Steinemann A. 2000. Rethinking human health impact assessment. *Environ Impact Assess Rev* 20:627–645.

Tiwari S, Dixit, S, Verma N. 2007. An effective means of biofiltration of heavy metal contaminated water bodies using aquatic weed *Echhornia crassipes. Environ Monit Assess* 129:253–256.

Ucer A, Uyanik A, Kutbay HG. 2013. Removal of heavy metals using *Myriophyllum verticillatum* (whorl-leaf watermilfoil) in a hydroponic system. *Ekoloji* 22(87):1–9.

Umavathi S, Longankumar K. 2010. Physicochemical and nutrient analysis of Singanallur Pond, Tamil Nadu (India). *Pollut Res* 29(2):223–229.

Valdman E, Erijman L, Pessoa FLP, Leite SGF. 2001. Continuous biosorption of Cu and Zn by immobilized waste biomass *Sargassum* sp. *Process Biochem* 36:869–873.

Van Dillewijn P et al. 2008. Bioremediation of 2,4,6-trinitrotoluene by bacterial nitroreductase expressing transgenic aspen. *Environ Sci Technol* 42:7405–7410.

Van Huysen T et al. 2003. Overexpression of cystathionine-gamma-synthase enhances selenium volatilization in *Brassica juncea. Planta* 218:71–78.

Vatamaniuk OK, Mari S, Lu YP, Rea PA. 1999. AtPCS1, a phytochelatin synthase from *Arabidopsis*: Isolation and in vitro reconstitution. *Proc Natl Acad Sci USA* 96:7110–7115.

Verhoeven JTA, Meuleman AFM. 1999. Wetlands for wastewater treatment: Opportunities and limitations. *Ecol Eng* 12(1):5–12.

Wcislo E, Ioven D, Kucharski R, Szdzuj J. 2002. Human health risk assessment case study: An abandoned metal smelter site in Poland. *Chemosphere* 47:507–515.

WHO (World Health Organization). 1997. Health and environment in sustainable development. Geneva: WHO.

WHO (World Health Organization). 2004. Manganese and its compound: Environmental aspects. Concise International Chemical Assessment Document 63. Geneva: WHO.

Wongsasuluk P, Chotpantarat S, Siriwong W, Robson M. 2014. Heavy metal contamination and human health risk assessment in drinking water from shallow groundwater wells in an agricultural area in Ubon Ratchathani province, Thailand. *Environ Geochem Health* 36:169–182.

Wylynko D. 1999. Prairie wetlands and carbon sequestration: Assessing sink under the Kyoto Protocol. Winnipeg, Manitoba, Canada: International Institute for Sustainable Development.

Xin K, Huang X, Hu JL, Li C, Yang XB, Arndt SK. 2014. Land use change impacts on heavy metal sedimentation in mangrove wetlands—A case study in Dongzhai Harbor of Hainan, China. *Wetlands* 34:1–8.

Yu C, Li H, Jia X, Chen B, Li Q, Zhang J. 2014. Heavy metal flows in multi-resource utilization of high-alumina coal fly ash: A substances flow analysis. *Clean Technol Environ Policy*. doi: 10.1007/s10098-014-0832-6.

Yuwono E, Jennerjahn TC, Nordhaus I, Ardli ER, Sastranegara MH, Pribadi R. 2007. Ecological status of Segara Anakan, Java, Indonesia, a mangrove-fringed lagoon affected by human activities. *Asian J Water Environ Pollut* 4:61–70.

Zenk MH. 1996. Heavy metal detoxification in higher plants—A review. *Gene* 179:21–30.

Zhang GS, Liu DY, Wu HF, Chen LF, Han QX. 2012. Heavy metal contamination in the marine organisms in Yantai coast, northern Yellow Sea of China. *Ecotoxicology* 21:1726–1733.

Zhang N, Zang S, Sun Q. 2014. Health risk assessment of heavy metals in the water of Zhalong wetland, China. *Ecotoxicology* 23:518–526.

Zhang Y, Liu J, Zhang J, Wang R. 2013. Energy-based evaluation of system sustainability and ecosystem value of a large-scale constructed wetland in North China. *Environ Monit Assess* 185:5595–5609.

Zhipeng H, Jinming S, Naixing Z, Peng Z, Yayan X. 2009. Variation characteristics and ecological risk of heavy metals in the south Yellow Sea surface sediments. *Environ Monit Assess* 157:515–528.

Zhu YL et al. 1999a. Overexpression of glutathione synthetase in Indian mustard enhances cadmium accumulation and tolerance. *Plant Physiol* 119:73–79.

Zhu YL et al. 1999b. Cadmium tolerance and accumulation in Indian mustard is enhanced by overexpressing gamma-glutamylcysteine synthetase. *Plant Physiol* 121:1169–1178.

Zietz BP, Dieter H, Lakomek M, Schneider H, Gaedtke KB, Dunkelberg H. 2003. Epidemiological investigation on chronic copper toxicity to children exposed via the public drinking water supply. *Sci Total Environ* 302:127–144.

4 Natural and Constructed Wetlands in Phytoremediation: A Global Perspective with Case Studies of Tropical and Temperate Countries

INTRODUCTION

The Millennium Ecosystem Assessment 2001–2005 (MEA) indicated that "the degradation and loss of wetlands is more rapid than that of other ecosystems," and that "the status of both freshwater and coastal wetland species is deteriorating faster than those of other ecosystems." This continuing degradation of natural wetlands necessitated the creation of diverse constructed wetlands to act as treatment wetlands for emerging contaminants derived from various anthropogenic activities.

Plants in a natural wetland provide a substrate (roots, stems, and leaves) upon which microorganisms can grow as they break down diverse emerging contaminants, like organic materials, and take up heavy metals (McCutcheon and Jørgensen, 2008). However, as a result of the exponentially increasing demands of human expansion and resource exploitation, it has been recognized that natural wetland ecosystems cannot always function efficiently for desired objectives, like the removal of recalcitrant emerging contaminants and the maintenance of stringent water quality standards (Wetzel, 1975, 1993). These and many other factors paved the way for the rapid development of constructed wetlands for waste (especially wastewater loaded with diverse xenobiotics and emerging contaminants) treatment (Wetzel, 1993; Yan et al., 2016; Rai et al., 2018). A constructed wetland is an artificial marsh or swamp that has been designed and constructed to utilize the natural processes involving wetland vegetation, soils, and their associated microbial assemblages to assist in waste and diverse emerging contaminant treatment (Song, 2003; Yan et al., 2016).

Constructed wetlands and treatment wetlands are engineered systems that have been designed to employ natural processes, including vegetation, soil, and microbial activity, to treat emerging contaminants present in a water environment. Further, constructed wetlands possess the merits of low cost and low- maintenance, and are capable of removing various emerging contaminants, including heavy metals, nutrients, organic matters, pharmaceuticals and personal care products (PPCPs), and micropollutants (Kadlec and Knight, 1996; Keffala and Ghrabi, 2005; Yan et al., 2016). In addition, constructed wetlands have recently been used for treating various emerging contaminants lying in wastewater and effluent types, including point source domestic sewage, acid mine drainage, agricultural wastewater, landfill leachate, and nonpoint-source storm water runoff (Kivaisi, 2001; Rousseau et al., 2008).

There has been much interest recently in the use of constructed wetlands for the removal of emerging contaminants (inorganic and organic) from contaminated soils, sediments, and waters (Horne, 2000a,b). In constructed wetlands, substrate interactions remove most emerging contaminants from wastewater, with plants serving as a "polishing system" (Matagi et al., 1998). The permanent or temporarily anoxic soils that characterize wetlands help to create conditions for the immobilization of emerging contaminants, like heavy metals, in the highly reduced sulfite or metallic form (Gambrell, 1994), while plants play an important role in metal removal via filtration, adsorption, and cation exchange, and through plant-induced chemical changes in the rhizosphere (Dunbabin and Bowmer, 1992; Wright and Otte, 1999). There is evidence that wetland plants can accumulate emerging contaminants, like inorganic heavy metals, in their tissues, such as duckweed (*Lemna minor*) (Zayed et al., 1998; Singh and Rai, 2016), water hyacinth (*Eichhornia crassipes*) (Vesk et al., 1999; Singh and Rai, 2016), salix (Stoltz and Greger, 2002), cattail (*Typha latifolia*), and common reed (*Phragmites australis*) (Ye et al., 2001; Yan et al., 2016). Cattail and common reed have been successfully used for the phytoremediation of Pb/Zn mine tailings under waterlogged conditions (Ye et al., 1997a,b).

The active reaction zone of constructed wetlands is the root zone (or rhizosphere). This is where physicochemical and biological processes take place for the phytoremoval of emerging contaminants and are induced by the interaction of plants, microorganisms, the soil, and pollutants (Stottmeister et al., 2003). Table 4.1 gives a concise overview of natural and constructed wetlands.

TABLE 4.1
Characteristic Features of Constructed and Natural Wetlands

Constructed Wetlands	Natural Wetlands
Man-made wetlands	Naturally exist
Selected organisms for a specific purpose	Multiple organisms with different ecosystems
Rely on natural wetland plants	Rely on themselves
Mostly smaller in area	Mostly large in area
Created from nonwetland sites	Created naturally
Most of them are utilized for the treatment of contaminated aquatic bodies	Multiple natural functions
Act as aesthetically pleasing wetlands	Most of them act as recreational places
Categories according to the treatment system applied	Include rivers, lakes, coastal lagoons, mangroves, peatlands, and coral reefs
Applicable to various types of wastewater for treatment, such as municipal, industrial, agriculture, and storm water	Nonapplicable

DIFFERENCE BETWEEN TERRESTRIAL AND WETLAND PHYTOREMEDIATION

Wetland phytoremediation is altogether different from terrestrial phytoremediation in many aspects. Domestic wastewater, agricultural runoff, urban storm runoff, acid mine drainage, and microbial pathogens are the categories of discharges or effluents that can only be treated through wetland systems, and terrestrial phytoremediation is not very relevant in this context. Further, although industrial wastewater treatment is possible with terrestrial phytoremediation, wetland biosystems can perform it better. In relation to emerging contaminants like heavy metals and some organics, the anoxic soils that usually characterize constructed wetlands tend to immobilize pollutants (and further retain them in nontoxic forms), which is noticeably different from terrestrial phytoremediation, where the oxidized soils tend to mobilize them into plant tissue. Moreover, constructed wetlands may be more efficient than natural wetlands, as they are designed for specific emerging contaminants and better suited to remediate them within stringent discharge limits. In conventional technologies, the remediation of emerging contaminants can be a high-energy affair, while in the case of constructed wetlands, the utilization of solar energy by wetland plants is extremely cost-effective.

CONSTRUCTED WETLANDS AND USE OF PHYTOREMEDIATION

Constructed wetlands are artificial wetlands planed and developed for the treatment process of contaminated water. They include some biotic and abiotic components found in natural wetlands, such as macrophytes, soil, microorganisms, and living organisms. The Environmental Protection Agency (EPA, 1993) states that constructed wetland systems (CWSs) are designed to mimic natural wetlands utilizing wetland plants and macrophytes, soils, and associated microorganisms to remove contaminants from wastewater effluents. The removal of emerging contaminants in these systems relies on a combination of physical, chemical, and biological processes that naturally occur in wetlands and are associated with vegetation, sediment, and their microbial communities (Farooqi et al., 2008). Large-scale constructed wetlands can successfully achieve the ecosystem functions of natural wetlands and hasten the restoration process, although the restoration effectiveness of ecosystem structures in terms of living biomass and water-using energy value is still inconclusive (Zhang Y et al., 2013). Phytoremediation and its use in constructed wetlands, as discussed earlier, is one of the effective cheaper eco-technologies, compared with advanced chemical technologies. These phyto-technologies promote sustainable use wetland plants and macrophytes in the treatment. The constructed wetlands technology is also applicable to a large amount of emerging contaminants from polluted water. This helps to minimize the concentrations of emerging contaminants. Constructed wetlands have been demonstrated to remove a significant percentage of emerging contaminants, like the heavy metals Cd, Cu, Pb, and Zn, from the effluent of road runoff over a period of 6 years (Gill et al., 2014). This system is applicable for all contaminated aquatic bodies, such as rivers, ponds, lakes, and estuaries.

Nowadays, constructed wetlands are engineered for effective phytoremediation processes using selective macrophytes. Engineered constructed wetlands or engineered wetlands are special, advanced, semi-passive kinds of constructed wetlands in which operating conditions are more actively monitored, manipulated, and controlled in such a manner as to allow emerging contaminant removal to be optimized (Anonymous, 2010). They are an aesthetically pleasing, solar-driven, passive technique that is useful for cleaning up wastes and emerging contaminants, including metals, pesticides, crude oil, polycyclic aromatic hydrocarbons (PAHs), landfill leaches (Zhang et al., 2010), and organic contaminants from mine waste, agricultural runoff, and industrial effluent (Williams, 2002). They have become an increasingly recognized pathway to advance the treatment capacity of wetland systems (Karim et al., 2004; Tsihrintzis et al., 2007; Kadlec and Wallace, 2008; Stefanakis et al., 2012; Zhang et al., 2012; Zhang BY et al., 2010; Yan et al., 2016).

WHY REPLACE NATURAL WETLANDS WITH CONSTRUCTED WETLANDS?

The excessive load of an expanding population and rapid industrialization has led to increased concentrations of several emerging contaminants. Therefore, mere dependence on natural wetlands did not suffice for the phytoremediation of emerging contaminants in natural wetlands. Further, climatic, environmental, and seasonal factors were limiting issues in natural wetlands, unlike designed constructed wetlands. In view of these factors and the diverse nature of emerging contaminants and pathogenic microbes, natural wetlands paved the way to constructed wetlands.

In natural wetlands, there is a short retention time of pollutant-loaded water in one place due to some existing flow, and moreover, the annual mass balance for nutrients in natural wetlands often shows that seasonal effects are also a constraint. Toxic and recalcitrant emerging contaminants can perturb the wetland ecosystem health by accumulating in wetland plants and causing the death of dependent fauna, like birds and fish. Interestingly, the most remarkable dissimilarity between constructed and natural wetlands is the isolation of the water regime from natural patterns. In natural wetlands, there is a seasonal alteration in the population dynamics of wetland plants due to abrupt changes in the water regime.

Constructed wetlands are man-made wetlands ecosystems that are similar to natural wetlands. The most attractive advantages of natural wetlands are their low energy requirement, straightforward operation, and maintenance works. After observations of the importance of natural wetlands, constructed wetlands are prepared to meet the needs of humans in a controlled way. They are designed to perform more functions than natural wetlands by choosing different selective materials. Most CWSs are used to clean up contaminated water through the process of phytoremediation, choosing an appropriate wetlands plant. They are created from a nonwetland ecosystem or a former terrestrial environment, mainly for the purpose of contaminant or pollutant removal from wastewater, as well as some emerging contaminants, like pharmaceuticals (Hammer, 1994; Yan et al., 2016). Compared with natural wetlands, more visits from tourists and less financial investment coming in as feedback into the wetland reduce system environment loading and promote self-support ability, ultimately generating sustainability (Zhang Y et al., 2013; Yan et al., 2016).

TYPES OF CONSTRUCTED WETLANDS

Constructed wetlands usually consist of a number of individual rectangular and/or irregularly shaped basins (cells) connected in series and surrounded by clay, rock, concrete, or other materials. Three types of cells may be used in a CWS: free water surface (FWS) cells, subsurface flow (SSF) cells, and hybrid cells, which incorporate surface and subsurface flows (http://nature-works.net/).

Constructed wetlands can be classified into several types based on their characteristics. One system defines four types based on the dominant plant species: (1) floating macrophytes (e.g., *Eichhornia crassipes* and *Lemna minor*), (2) floating-leaf macrophytes (e.g., *Nymphea alba* nad *Potamogeton gramineus*), (3) submersed macrophytes (e.g., *Littorella uniflora* and *Potamogeton crispus*), and (4) emerged rooted macrophytes (e.g., *Phragmites australis* nad *Typha latifolia*) (Arias and Brix, 2003).

(a)

(b)   (c)

FIGURE 4.1 Wetland plants used globally for the remediation of emerging contaminants in several constructed wetlands. (Note: *Eichhornia crassipes, Lemna minor, Salvinia cucullata,* and *Pistia stratiotes* were used during our microcosm phytoremediation experiment for the removal of Fe.) (a) *Typha latifolia.* (b) *Phragmites karka.* (c) *Arundo donax.*   *(Continued)*

FIGURE 4.1 (CONTINUED)   Wetland plants used globally for the remediation of emerging contaminants in several constructed wetlands. (Note: *Eichhornia crassipes, Lemna minor, Salvinia cucullata,* and *Pistia stratiotes* were used during our microcosm phytoremediation experiment for the removal of Fe.) (d) Global wetland plants of a floating nature: *Eichhornia crassipes, Lemna minor,* and *Salvinia cucullata.* (e) *Pistia stratiotes* inside an aquarium. (From Singh, M.M., and Rai, P.K., *Int. J. Phytoremediation,* 18(12), 1231–1236, 2016.) (f) *Juncus* sp. (g) *Azolla* sp.

Figure 4.1a–d shows diagrammatic representations of certain common global wetland plants of constructed wetlands.

There are two types of constructed wetlands, which are classified according to wetlands hydrology:

1. Free water surface: They are similar to wetlands that have a shallow flow of water over-saturated substrate.
2. Subsurface flow: In this system, the water flows horizontally or vertically through the substrate, which supports the growth of plants (Wu et al., 2015).

Another common classification of constructed wetlands is (1) surface flow wetlands, (2) horizontal SSF wetlands, (3) vertical SSF wetlands, and (4) hybrid systems

(Arias and Brix, 2003). (Refer to Chapter 5 for figure or diagrammatic representations of these wetland types.)

ROLE OF PLANTS IN CONSTRUCTED WETLANDS

It is well known that soil, sediments, hydrology, and vegetation are integral components of constructed wetlands. Plants play a major role in the structural and functional aspects of aquatic ecosystems. They are the most important part of constructed wetlands. Constructed wetlands are functionless without plants because they perform the cleanup process. Plants have a natural ability to take up, accumulate, or degrade emerging contaminants, whether organic and inorganic substances (Lasat, 2000) or heavy metals, through the process of bioaccumulation (Tiwari et al., 2007). Various species show different behavior regarding their ability to accumulate in roots, stems, and/or leaves (Kumar et al., 2008; Yan et al., 2016).

Wetland macrophytes can be classified based on their physiology, morphology, and submerged or emergent type. They also have certain anatomical adaptations in macrophytes to cope with wetland environments, for example, root pockets and swollen leaf base in water hyacinth (*Eichhornia crassipes*) (Figure 4.1a–g).

Free-floating type: This type usually floats on the surface of water and is not rooted to the substrate. Water hyacinth (*Eichhornia crassipes*), water fern (*Azolla pinnata*), and duckweeds (five major genera: *Lemna*, *Spirodela*, *Wolffia*, *Wolffiella*, and *Landoltia*) are popular wetland plants. They are more frequent in global constructed wetlands of tropical and temperate countries. They are efficient plants for the eco-removal of emerging contaminants in global wetland systems. *Pistia stratiotes*, *Salvinia herzogii*, *Wolffia columbiana*, *Lemna valdiviana*, *Nymphaea* spp., *Nuphar advena*, and *Juncus effusus* are some of the popular free-floating types of wetland plants.

Submerged type: The submerged type either floats in the photic region of water or is rooted in the substrate, or both situations may be applicable, for example, *Myriophyllum spicatum*, *Heteranthera dubia*, *Hydrilla*, *Vallisneria*, and *Ceratophyllum*. They are less common in global constructed wetlands of tropical and temperate countries.

Emergent type: These wetland plants, like giant reed, bulrush, common cattail, and common reed, are more useful in the constructed wetlands of tropical and temperate countries and are efficient in the eco-removal and phytoremediation of emerging contaminants. *Phragmites australis*, *Phalaris arundinacea*, *Typha domingensis*, *Typha latifolia*, *Phragmites karka*, *Juncus pallidus*, and *Empodisma minus* are some popular emergent wetland plants.

Table 4.2 represents the remarkable benefits of plant-based treatment wetlands.

TABLE 4.2

Summary of Known Uses of Phytoremediation and Treatment Wetlands

Emerging Contaminants and Water Quality Parameters of Remediated Wetland Systems	Human Problems and Health Concerns	Environmental Problems and Aesthetic Concerns
BOD (water quality parameter) A unique parameter that is considered to be an indicator of aquatic ecosystem health; the higher the BOD value, the greater the intensity of water pollution	Drinking water quality is being perturbed, malodors reduce the aesthetic pleasure; health hazards are indirectly correlated with BOD increase	Fish kills from reduced dissolved oxygen (DO) due to algal (*Microcystis*, a cyanobacteria) blooms
Nitrate (water quality parameter) A water quality parameter of paramount importance due to the N requirements of wetlands and aquatic organisms and its intimate linkage with nutrient enrichment or eutrophication perturbing the health of wetland systems	Blue baby disease—a dreaded disease caused by excess nitrate, affecting lake use	Eutrophication (nitrate nutrient, phosphate enrichment); avian botulism is also a common occurrence
Nanoparticle (NP) particulates	Lake use drastically affected by an excess load of NP particulates; interestingly, the sources of particulates may be both air and water, adversely affecting the turbidity and hence light penetration inside water columns	Water clarity drastically affected, and as stated, light penetration can further alter the metabolic machinery
Phosphorous (water quality parameter) May also be a limiting factor for the phytoplankton community in wetlands and is intimately linked with nutrient enrichment or eutrophication perturbing the health of wetland systems (like in the nitrate case)	Lake use affected due to the dominance of specific life-forms that overutilize the dissolved oxygen (DO), leading to loss of wetland biodiversity in totality, in addition to reducing aesthetic pleasure	Eutrophication (nitrate nutrient, phosphate enrichment)

(Continued)

TABLE 4.2 (CONTINUED)
Summary of Known Uses of Phytoremediation and Treatment Wetlands

Emerging Contaminants and Water Quality Parameters of Remediated Wetland Systems	Human Problems and Health Concerns	Environmental Problems and Aesthetic Concerns
Heavy metals (Cu, Pb, acid mine drainage, storm runoff)	Drinking water standards are perturbed	Toxicity to flora and fauna of wetland systems
Metalloids (Se from agriculture, copiers, taillight production) Emerging contaminant	Toxicity to livestock and other wetland organisms	Bird embryo deformities, skeletal deformation in fish of wetland systems
Pesticides	Food chain or web toxicity, resulting in cancers; biomagnification is the basic mechanism in operation—there are successive exponential increases in the concentrations of these organic emerging contaminants as we move up trophic levels, and humans, at the top of the trophic levels, are the ultimate victims through intake of contaminated fish	Nontarget organism deaths in wetland systems
Trace organics (chlorinated organics, estrogen mimics, PPCPs) Emerging contaminants (organics)	Major long-term objections to human water reuse due to resulting health hazards	Subtle toxic effects in organisms of wetland systems
Bacterial and microbial pathogens	Microbial pathogenic diseases due to contaminated drinking water	Not applicable

Source: Modified after Horne (2000a,b).

MECHANISM OF PHYTOREMEDIATION
IN CONSTRUCTED WETLANDS

In Chapters 2 and 3, we discussed the mechanism of phytoremediation in detail. Here, we represent only glimpses of it to maintain connectivity. It is well known that the huge cost burden imposed from chemical technologies has opened a path in the marketplace for innovative and alternative technology, that is, phytotechnologies in the form of phytoremediation. Phytoremediation technology is of particular relevance in emerging contaminant removal in constructed wetlands and comprises the following mechanisms (Rai, 2009) (Figure 4.2):

- Phytoextraction, in which metal-accumulating plants are used to transport and concentrate metals into the harvestable parts of roots and aboveground shoots
- Rhizofiltration, in which plant roots absorb, precipitate, and concentrate toxic metals from polluted effluents
- Phytostabilization, in which the mobility of heavy metals is reduced through the use of tolerant plants
- Phytotransformation and phytodegradation, in which a contaminant can be eliminated via phytodegradation or phytotransformation by plant enzymes or enzyme cofactors

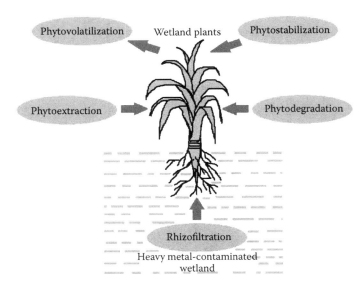

FIGURE 4.2 Phytoremediation mechanisms and steps in wetlands for diverse emerging contaminants, like heavy metals.

PHYSICAL EFFECTS OF ROOT STRUCTURE OF WETLAND PLANTS IN CONSTRUCTED WETLANDS

Several authors have argued that the most important mechanism by which plants contribute to the constructed wetland treatment process is not in uptake, but rather in the physical effects of their root structure combined with aeration (Tanner, 2001; Brisson and Chazarenc, 2009). Root growth is well known to affect some soil hydraulic qualities (Cooper and Boon, 1987; Stottmeister et al., 2003). The physical effects of roots include filtering, flow velocity reduction, improved sedimentation, decreased resuspension, and even the distribution of water and prevention of clogging (Stottmeister et al., 2003; Wiessner, 2003; Vymazal, 2011b).

PLANT–MICROBE INTERACTION IN CONSTRUCTED WETLANDS

It is well known that the active reaction zone of constructed wetlands is the root zone (or rhizosphere). Further, the term *rhizosphere* can be subdivided into the endorhizosphere (the root interior) and the ectorhizosphere (the root's surroundings). The zone in which these two areas meet is known as the rhizoplane (Stottmeister et al., 2003). This is where the most intense interaction between the plant and microorganisms is expected for the removal of emerging contaminants, and it is rightly regarded as the black box of wetland plants (Stottmeister et al., 2003).

Plant roots are inextricably associated with microorganisms called endophytes. They include endophytic bacteria, endophytic fungi, and endophytic ectinomycetes (Raghukumar, 2008). Their activity inside the plants helps increase their metabolism, forming a beneficial association between them. Interestingly, endophytes help plants enhance growth through phytohormone production; supplying nitrogen after the nitrogen fixation process; making them resistant to environmental stresses (heat, cold, drought, and salt); producing important medicinal, agricultural, and industrial compounds; and enhancing phytoremediation after improving the uptake of contaminants and degradation of several toxins (Khan and Doty, 2011). In the phytoremediation of emerging contaminants, microbial cells are controlled by multiple mechanisms and interactions, including adsorption, chelation, entrapment, and ion exchange (Gadd, 2004).

It is worth mentioning that wetland plants or macrophytes provide the litter layer that provides both the microbial habitat and a source of labile carbon for bacterial and microbial processes involved in the eco-removal of emerging contaminants (Horne, 2000b). Further, rhizospheric bacteria or other beneficial microbes play an important role in the uptake of emerging contaminants, particularly heavy metals, by wetland plants or macrophytes (Weis and Weis, 2004). Further, De Souza et al. (1999) observed that when bacteria were inhibited with antibiotics, plants (*Scirpus robustus* and *Polypogon monspeliensis*) accumulated lower concentrations of selenium (Se) and mercury (Hg).

Mycorrhizae (symbiotic fungi associated with roots of angiospermic and higher plants) provide an interface between the roots and the soil, increasing the absorptive surface area of root hairs, and are effective at assimilating emerging contaminants like metals that may be present at toxic concentrations in the soil (Meharg and Cairney, 2000; Weis and Weis, 2004). In this context, Khan et al. (2000) suggested that mycorrhizae play a protective role, restricting the uptake of metals by plants by immobilizing the emerging contaminants, specifically metals in the fungal tissue. Nevertheless, there is paucity of studies on the symbiotic role of microbes in emerging contaminants or metal uptake, and future studies should be specifically focused in this direction.

Marchand et al. (2010) proposed a new index, the relative treatment efficiency index (RTEI), to quantify treatment impacts on emerging contaminants like metal removal in constructed wetlands. They emphasized that further research is needed on key components, such as the effects of differences in plant ecotypes and microbial communities, in order to enhance metal removal efficiency.

Rhizospheric endophytes, as explained above, are densely colonized inside the plant roots, decreasing from the stem to the leaves (Porteous-Moore et al., 2006). In comparison with other rhizosphere microbes, endophytes interact more closely with their host plants and could more efficiently improve phytoremediation (Zhang et al., 2011). Among the endophytic genera, Burkholderiaceae, Enterobacteriaceae, and Pseudomonaceae are the most common cultivable species found in constructed wetlands to interact with emerging contaminants (Khan and Doty, 2011).

These microbes are treated through plants in the constructed wetlands to enhance the various cleanup processes. Brix (1997) and Vymazal (2011b) pointed out the significance of the rhizosphere in creating improved conditions for various microbes in constructed wetlands. Since microbes are considered key drivers in the treatment process, any factor that changes their composition, biodegradation efficiency, or concentration has a significant impact on the whole constructed wetland.

In addition to the microbial mode of action in the rhizosphere, the root release of liquid exudates (consisting of metaboloites and enzymes) and gas is likely a key component of plant effects in constructed wetlands. The extensive work by Armstrong and others revealed various aspects of oxygen release by plant roots (Drew et al., 1985; Armstrong and Armstrong, 1990; Armstrong et al., 2009). Root oxygenation occurs in the daylight and depends to some extent on photosynthetic activity (Stein and Hook, 2005; Williams et al., 2010). Oxygenation by roots has been shown to have a significant impact on important mechanisms of wastewater treatment in constructed wetlands, including influence on redox potential (Bialowiec et al., 2012), which is critical in determining the nitrogen fate, oxidation of some phytotoxins (Armstrong et al., 2000), and enhancement of microbial activity (Ueckert et al., 1990).

PLANT UPTAKE OF CONTAMINANTS
IN CONSTRUCTED WETLANDS: GLOBAL STUDIES

By utilizing N, P, and other nutrients, plants can reduce the concentrations of elements that would otherwise be considered pollutants in constructed wetlands. Plants can also accumulate phytotoxic elements, such as heavy metals and other emerging contaminants, in vacuolar or granular compartments. Thus, phytoremediation may be an important role for wetland plants in constructed wetlands for the eco-removal of emerging contaminants. Nevertheless, several researchers have found that plant uptake in constructed wetlands is negligible (Brix, 1994; Geller, 1997; Brisson and Chazarenc, 2009; Langergraber and Simunek, 2005), while others have claimed that it is significant. Another important aspect of plant uptake is the effect of plants on the water balance in constructed wetlands.

EVAPOTRANSPIRATION

Wetlands receive water through influx and rain, and lose water to outflow and evaporation. Wetland plants have a critical role in determining the dynamics of water loss, mainly by dictating water loss through evaporation and plant transpiration, that is, evapotranspiration (ET). The ET of emergent macrophytes is a significant process in constructed wetlands (Wetzel, 1975) and may reach high levels—seven to eight times higher than actual evaporation without plants. For example, ET from a constructed wetland in Morocco planted with *Arundo donax* was 40 mm/day and nearly 60 mm/day with *Phragmites australis*, compared with 7 mm/day in an unplanted horizontal flow constructed wetland (El Hamouri et al., 2007). Borin et al. (2011) found a similar amplification of water loss from a constructed wetland planted with *P. australis* in a rather humid area in Italy. Water loss from the constructed wetland through ET slows flow velocities, which induces longer retention times, and increases pollutant and salt concentrations in the water (Headley et al., 2012). The transfer of water to the atmosphere is sometimes an advantage, mainly in humid environments (Gregersen and Brix, 2001). On the other hand, in arid regions, where treated wastewater loaded with emerging contaminants is destined for reuse, water loss may be disadvantageous (Green et al., 2006; El Hamouri et al., 2007; Masi and Martinuzzi, 2007). Nevertheless, data on ET rates in constructed wetlands in general, and in arid lands in particular, are still scarce (Tencer et al., 2009; Borin et al., 2011).

MICROCLIMATIC CONDITIONS

The physical structure of wetland plants growing in the constructed wetland medium affects the microclimatic conditions in the system, which may then have a significant effect on various components of the system. The effect of plants on the microenvironmental conditions of the constructed wetland includes shade that prevents algal growth, and insulation from radiation in the spring and frost in the winter. Reduced wind velocity by the plants' upper parts may stabilize the sediment surface.

Other functions of wetland plants in the constructed wetland include aesthetic appearance, the elimination of microbial pathogens and insects, and offensive odors. However, some plant functions are negative. Knight et al. (1999) reported enhanced mosquito reproduction in the presence of plants.

PLANT PRODUCTION IN CONSTRUCTED WETLANDS

Constructed wetland plants can be used for the production of marketable goods. The literature includes reports on growing vegetation for fibers to be used in construction material, like *Typha* spp. (Maddison et al., 2009), and bioenergy crops, as in the case of short-rotation willow coppice harvest in constructed wetlands (Aronsson and Perttu, 2001). Some plants are produced in constructed wetlands as submerged ornamental plants, or conversely, some ornamental plants have been examined as constructed wetland plants, to enable wastewater treatment with the additional benefit of commercial value (Belmont and Metcalfe, 2003). It is worth testing other potential products, such as water lilies and animal feed.

CONSTRUCTED WETLAND PROCESSES FACTORS

Constructed wetland processes can be enhanced by optimizing certain factors. Some important factors include:

1. Constructed wetland depth: Constructed wetland depth is very important for determining the plant types (emergent, submerged, and free-floating wetland plants) that are required for the treatment of emerging contaminants in the phytoremediation processes and also affect the redox status and level of dissolved oxygen (Dwire et al., 2006).
2. Hydraulic loading rate (HLR) and hydraulic retention time (HRT): The efficiency of the constructed wetlands depends on the optimal design of HLR and HRT to eco-remove emerging contaminants in the phytoremediation processes. Interestingly, the greater the HLR, the quicker the passage of the water through the media (Wu et al., 2015). Longer HRTs let the plants eco-remove emerging contaminants in higher amounts after the establishment of an appropriate microbe (Huang et al., 2000).

ECO-REMOVAL OF EMERGING CONTAMINANTS AND METAL ACCUMULATION BY PLANTS

The uptake and accumulation of emerging contaminants directly from water bodies and assimilation by plants are the greatest benefit of phytoremediation (Yang et al., 2005; Rai, 2009). As discussed in earlier chapters, phytoremediation is a low-cost, sustainable, and green remediation strategy for cleansing emerging contaminants like heavy metals in contaminated landscapes (Arthur et al., 2005; Rai, 2007a,b, 2008a–c, 2009, 2010a–d, 2011, 2012, 2016; Rai et al., 2010, 2017; Singh and Rai, 2016; Rai and Singh, 2016). Interestingly, for normal growth and metabolism, plants require some contaminants like metals (e.g., Cu, cobalt [Co], Fe, molybdenum [Mo], Mn, Ni, and Zn) are essential in small amounts (Hänsch and Mendel, 2009) and a few others (e.g., As, Cd, Hg, and Se) in traces (Rascio and Navari-Izzo, 2011). Nevertheless, the concentration factors in the total environment make them emerging contaminants and pollutants. Moreover, wetland plants require Cu to mediate CO_2 assimilation; Cu is essential in electron transport mechanisms in photosynthesis and respiration (Demirevska-Kepova et al., 2004). However, when these emerging contaminants and metals occur at more than optimal levels, as in contaminated land and surface and groundwater sites, plants suffer toxicity, resulting in stressed conditions. Stress, including oxidative stress, from emerging contaminants manifests as reduced vigor and growth inhibition in diverse wetland plants (Levitt, 1980; Stadtman and Oliver, 1991). Excess concentrations of Cu and Cd in the substratum affect germination, seedling growth, and lateral root production in *Solanum melongena* (Solanaceae) (Neelima and Reddy, 2004). Excess Mn in leaves reduces photosynthetic efficiency (Kitao et al., 1997). Mn toxicity manifested as necrotic spots on leaves, petioles, and stems of *Glycine max* (Fabaceae) (Wu, 1994).

Wetland plants used in metal- and emerging contaminant–inundated sites can either indicate, exclude, or accumulate them. Accumulation capability builds on the genetically driven physiological strengths of plants in absorbing and storing emerging contaminants from the sediment and water (Knight et al., 1997). Detoxification of emerging contaminants like metal ions is the principal function. Plants such as *Festuca rubra* cv. Merlin (Poaceae) are valued for their capability to exclude and stabilize erosion-prone metal-contaminated soils (Wong et al., 1994). To this end, *Thlaspi caerulescens* (Brassicaceae) and *Viola calaminaria* (Violaceae) are examples of established accumulator species, which do not exclude metals from entering the root, but they detoxify high metal levels upon accumulation in cells (Baker, 1981). The excluder species retain the absorbed emerging contaminants, for example, metals in their roots, by altering membrane permeability, changing the binding capacity of the tonoplast and cell walls, and chelating the metal with a ligand, for example, citrate and malate (Lasat, 2000). The accumulator species transport and retain metals in their shoot tissues proportional to metal levels occurring in the substratum (Ghosh and Singh, 2005). In constructed wetlands, to avoid the limitation of large land use, a biohedge water hyacinth wetland is frequently used for the remediation of diverse emerging contaminants. Sand filtration may be a very useful mechanism to eliminate emerging contaminants in general and pathogenic microbes in particular (Diaz, 2016). In China, the interspecific competition of a couple of popular wetland plants, *Phragmites* and *Typha*, was investigated in two large constructed wetlands (FWS and horizontal SSF systems), and the results revealed the superiority of *P. australis* (Zheng et al., 2016).

The hyperaccumulators—wetland plants that can store large quantities of emerging contaminants like metals—store metals in shoots (up to >1.0% of Zn and Mn; up to 0.1% of aluminum [Al], As, Se, Ni, cobalt, Cr, Cu, and Pb; and up to 0.01% of Cd) (Baker and Brooks, 1989). The capability to hyperaccumulate metals is known in about 450 species of flowering plants (Sarma, 2011), which take up, transport, and sequester emerging contaminants, like antimony (Sb), As, Cd, Co, Cu, Mn, Pb, Se, titanium (Ti), and Zn metals. Hyperaccumulating wetland plants differ from excluder species by actively taking up and translocating large volumes of either one or more metals from roots to aerial organs, particularly the leaves, at concentrations 100–1000 times greater than those known in nonhyperaccumulating, excluding species (Rascio and Navari-Izzo, 2011). Wetland plants are useful in bringing both contaminated terrestrial and wetland systems within regulated limits.

Eichhornia crassipes and *Pistia stratiotes* were demonstrated to reduce the toxicity of wetlands emanating from emerging contaminants, and thus reduced fish mortality and kept the ecosystem services and aesthetic pleasure intact (Victor et al., 2016). Moreover, in the context of the elimination or amelioration of pollutants, *Canna indica* and common reed were wetland plants efficient in the eco-removal or phytoremediation of emerging contaminants (Cu, Zn, carbamazepine, and linear alkylbenzene sulfonates in their tissues [Macci et al., 2014]). Duckweeds were found to be efficient in the phytoremediation of selenium (Mechora et al., 2014). Further, wetland plants from the duckweed family are a small group of flowering plants, like *Lemna*, *Spirodela*, *Wolffia*, *Wolffiella*, and *Landoltia*, that proved their worth in the eco-removal or phytoremediation of nutrients and emerging contaminants (PPCPs, pesticides, and surfactants) from natural, domestic, industrial, and agricultural effluents. However, more global research at a lab scale and more future studies on a pilot or field scale are warranted (Forni and Tommasi, 2015).

ECO-REMEDIATION OF EMERGING CONTAMINANTS AND HEAVY METAL REMOVAL PROCESSES IN WETLANDS: MECHANISMS

Four mechanisms affect emerging contaminant and metal removal in wetlands (Lesage et al., 2007): (1) adsorption to fine-textured sediments and organic matter (Gambrell, 1994), (2) precipitation as insoluble salts (mainly sulfides and oxyhydroxides in the form of emerging contaminants), (3) absorption and induced changes in biogeochemical cycles by plants and bacteria (Kadlec and Knight, 1996), and (4) deposition of suspended solids due to low flow rates. All these reactions lead to the accumulation of emerging contaminants and metals in the substrate of wetlands. The efficiency of systems depends strongly on the (1) inlet metal and emerging contaminant concentrations and (2) hydraulic loading (Kadlec and Knight, 1996).

Thlaspi caerulescens, *Nicotiana tabacum* (Solanaceae), and *Zea mays* (Poaceae) (Kayser et al., 2000) possess heavy metal–sequestering capabilities, such as binding the emerging contaminants and metal ions to their cell walls, reducing the efflux pumping of metal ions at the plasma membrane, chelating metal ions in the cytosol with different ligands (e.g., phytochelatins, metallothionein (MT), and metal binding proteins), and sequestering emerging contaminants like metal ions in vacuoles enabled by transporters in the tonoplast (Rai, 2009; Adams et al., 2013). Proton pumps of the plasma membrane are also valuable tool to chelate the emerging contaminants and metals (Mukhopadhyay and Maiti, 2010; Adams et al., 2013).

Sorption (comprising physisorption, i.e., physical processes with weak bindings, and chemisorption, i.e., chemical processes with strong bindings of emerging contaminants), the transfer of ions from a soluble phase to a solid phase, is an important mechanism for the removal of emerging contaminants and metals in wetlands. Sorption may result in short-term retention or long-term stabilization of emerging contaminants. Sorption describes a group of processes, which includes adsorption, absorption (e.g., with biochemical processes when an emerging contaminant or compound from the external medium is entering into a biosystem), and precipitation reactions. Metals and emerging contaminants are adsorbed to particles by either ion exchange, depending on factors such as the type of element and the presence of other elements competing for adsorption sites (Seo et al., 2008), or chemisorption. The retention of emerging contaminants like Pb, Cu, and Cr by adsorption is greater than that of Zn, Ni, and Cd in constructed wetlands (Sheoran and Sheoran, 2006).

Some emerging contaminants, for example, metals like Fe, Al, and Mn, can form insoluble compounds through hydrolysis and/or oxidation. This leads to the formation of a range of oxides, oxyhydroxides, and hydroxides of emerging contaminants (Sheoran and Sheoran, 2006). Fe removal depends on pH, redox potential, and the presence of anions (ITRC, 2003). The amounts and forms of Fe in solution strongly affect metal and emerging contaminant removal. Fe(II) is soluble and represents an important bioavailable fraction. It can be oxidized to Fe(III) in conjunction with H$^+$ ion consumption under aerobic conditions (Jonsson et al., 2006). It is worth mentioning that Fe(III) can deposit onto root surfaces of aquatic macrophytes or wetland plants (Weiss et al., 2003), forming *plaques* in dicots and *siderophores* in monocots with a large capacity to adsorb metals (Doyle and Otte, 1997; Cambrolle et al., 2008), aided also by the action of Fe(II)-oxidizing bacteria (Emerson et al., 1999). Furthermore, Fe(III) can precipitate to produce oxides, hydroxides, and oxyhydroxides with which other metals may coprecipitate. Likewise, Fe(II) can also precipitate as oxides (Jonsson et al., 2006) or coprecipitate with other metals, such as Zn, Cd, Cu, or Ni (Matagi et al., 1998). Iron oxides have a particularly strong affinity for cations with a similar size compared with Fe(III) and Fe(II), for example, Zn, Cd, Cu, and Ni (Dorman et al., 2009). Therefore, those emerging contaminants may combine with Fe, forming metal oxide complexes (Benjamin and Leckie, 1981). This coprecipitation is limited when Fe(II) forms complexes with, for example, SO_4^{2-} (Sung and Morgan, 1980), thus reducing the potential of metal removal. Importantly, arsenic, a metalloid of wide human health concern, may also be removed from the water column by adsorbing onto amorphous iron hydroxides or by coprecipitating with iron oxyhydroxides (Manning et al., 1998).

In a constructed wetland of China, *Cyperus alternifolius* interaction with pharmaceutically active compounds (PhACs) has been demonstrated effectively and systematically (Yan et al., 2016). In constructed wetlands in Poland, a *Salix* wetland system was effective in purification overall, while a *Populus* wetland system was most effective for Cu and Ni (Samecka-Cymerman et al., 2004). However, in this constructed wetland of Poland, a *Phragmites* system was most effective in purification and removal.

A constructed wetland based on a matrix with exclusively reducing conditions, however, cannot be efficient for diverse emerging contaminants. These conditions promote massive ion release, particularly of Fe and Mn, into the water by reduction of the oxides and oxyhydroxides trapped in the substrate (Goulet and Pick, 2001). Most emerging contaminants, particularly metals in the pore water, precipitate as metal oxides or adsorb onto organic matter at redox potentials higher than 100 mV. Between 100 mV and −100 mV, metal oxides are reduced, resulting in a release of dissolved metals. These metals can still adsorb onto organic matter if adsorption sites are available. Most macrophytes or wetland plants play a significant role in maintaining oxidizing conditions by shoot-to-root oxygen transport (Armstrong, 1978). Such conditions promote the formation of iron oxides, hydroxides, and oxyhydroxides, such as the iron plaques and siderophores, and consequently result in metal and other emerging contaminant removal by adsorption and coprecipitation.

Heavy metals may also form metal carbonates. Although carbonates are less stable than sulfides, they can contribute to the initial trapping of metals, an emerging contaminant of concern (Sheoran and Sheoran, 2006). Interestingly, carbonate precipitation is especially effective for the removal of Pb and Ni (Lin, 1995). According to Maine et al. (2006), the incoming wastewater composition containing high pH, carbonate, and calcium concentrations favored the metal retention in the sediment. Calcium carbonate precipitation represents an important pathway governed by the incoming water pH as metals, an emerging contaminant of concern, are removed or adsorbed onto carbonates.

The water quality parameter of wetland systems, like pH, strongly affects the efficiency of metal or other emerging contaminant removal in wetlands. Ammonium conversion into nitrites during nitrification leads to proton production. These hydrogen ions are then neutralized by bicarbonate ions. Macrophytes or wetland plants, in releasing oxygen, promote the nitrification process. Further, protons produced due to nitrification may not all be neutralized by HCO_3^- ions, resulting in a pH decrease in wetland systems (Lee and Scholz, 2007). The overall mean surface charge of ferric oxyhydroxides changes from a positive to a negative value as pH increases. Hence, to promote the adsorption and removal of oxyanions of, for example, As, Sb, and Se, iron coprecipitation must occur under acidic conditions (Sheoran and Sheoran, 2006). Conversely, alkaline conditions are necessary to promote the coprecipitation of cationic metals, such as Cu, Zn, Ni, and Cd. A high rate of nitrification can therefore reduce the efficiency of a constructed wetland in terms of cationic metal removal (Lee and Scholz, 2007).

In the special case of acid mine drainage due to emerging contaminants of low pH, the water and substrates are characterized by high metal concentrations and a low pH. When sulfide minerals contained in mine drainage are exposed to atmospheric and dissolved oxygen, they are oxidized (Holmstrom, 2000). For example, pyrite (FeS_2) is oxidized to form soluble Fe(II), SO_4^{2-}, and H^+. This leads to the release of dissolved iron and protons, which in turn leads to the release of other metallic ions, such as Mn, Ni, Zn, Cu, and Cd. Because of the extreme conditions of acid mine drainage, wetlands may have a low potential for water treatment (Nyquist and Greger, 2009). Yet, to date more than a thousand wetland systems have been constructed specifically for the eco-removal of emerging contaminants from mine drainage (Skousen and Ziemkiewicz, 1995). The formation of acid mine drainage can be prevented by limiting contact between mining wastes and oxygen in the total environment. One attractive and efficient solution for reducing O_2 diffusion is to construct a wetland as a cover over mine waste (Stoltz and Greger, 2005).

Macrophytes or wetland plants, such as *Phragmites australis*, promote the sedimentation of suspended solids and prevent erosion by decreasing water flow rates by increasing the length per surface area of the hydraulic pathways through the system (Lee and Scholz, 2007). Surface flow systems may exhibit either static behavior, in which there is virtually no flow, or dynamic behavior, in which water is passing through at relatively high flow rates. Under static conditions, the wetland systems behave like a stagnant pond in which displacement effects caused by submerged plant mass decrease retention times. Under dynamic conditions, active flow-through and stem drag are increased and more important than displacement of volume. Retention times increase with increasing vegetation density, thus enabling better sedimentation. For particles less dense than water, sedimentation is possible only after floc formation. Flocs may adsorb other types of suspended materials, including metals.

Flocculation is enhanced by high pH, high concentrations of suspended matter, high ionic strength, and high algal densities (Matagi et al., 1998).

The aforesaid discussion proved the utility of wetland plants to eco-remove emerging contaminants. Nevertheless, more research is warranted toward certain emerging contaminants, like imazalil and tebuconazole, two commonly used systemic pesticides, as four wetland plants, *Typha latifolia*, *Phragmites australis*, *Iris pseudacorus*, and *Juncus effusus*, had very low removal (2.8%–14.4%) of their spiked concentrations (Lv et al., 2016).

ECO-REMOVAL OF EMERGING CONTAMINANTS AND WATER QUALITY PARAMETERS BY CONSTRUCTED WETLANDS IN TROPICAL AND TEMPERATE COUNTRIES: GLOBAL CASE STUDIES

Wetland systems and other aquatic ecosystems of China are mostly contaminated by heavy metals and other emerging contaminants. In China, there are numerous metal mines and the process of metal mining has caused severe heavy metal pollution (Deng et al., 2004). Xiao-Bin et al. (2007) used ornamental hydrophytes and wetland plants (*Iris pseudacorus* L. and *Acorus gramineus*) in constructed wetlands to treat urban or rural domestic wastewater, while *Leersia hexandra*, *Juncus effuses*, and *Equisetum ramosisti* performed well in field conditions of China (Deng et al., 2004). Zhang et al. (2010), in their review, mentioned the utility of constructed wetlands in a case study of full-scale application in Newfoundland, Canada. The constructed wetland in Flanders, Belgium, has increased more than a hundred in numbers during the last 10 years (Lesage et al., 2007). The role and keen selection of macrophytes and wetland plants for constructed wetlands was demonstrated through a microcosm investigation in France (Guittonny-Philippe et al., 2015). *Phragmites australis* in a constructed wetland of north Italy was found to be efficient in metal phytoremediation. Samecka-Cymerman et al. (2004) demonstrated metal removal in a constructed wetland of Poland. Other wetland plants, like *Limnocharis flava*, *Thalia geniculata*, and *Typha latifolia*, were useful in metal phytoremediation in constructed wetlands of Ghana, and the treated water may be used for irrigating the agricultural fields (Anning et al., 2013). Coastal wetlands in Ghana also removed metals (Gbogbo and Otoo, 2015). The performance efficiency of constructed wetlands was appreciable with *Azolla pinnata* in the wastewater treatment generated from the Araromi community in Akure, Nigeria (Akinbile et al., 2016), while *Phragmites karka* and *Veteveria nigritana* were demonstrated to remediate the metals in constructed wetlands of a steel manufacturing company in southwest Nigeria (Badejo et al., 2015). Likewise, constructed wetlands were also demonstrated to be useful in Pakistan pertaining to metal phytoremediation. Similarly, in Pennsylvania a constructed wetland based on emergent *T. latifolia* was useful in the metal treatment of a coal mine leachate (Ye et al., 2001). Moreover, water hyacinth (*Eichhornia crassipes*) was a potential biosystem for metal phytoremediation in Taiwan and other countries of the Asia-Pacific region (Liao and Chang, 2004; Rai and Singh, 2016; Singh and Rai, 2016). Similarly, *T. latifolia* and *P. australis* were useful in the metal phytoremediation of an engineered constructed wetland in Taiwan. To this end, Turker et al. (2013) demonstrated boron (a macronutrient in plants) removal in a constructed wetland of

Turkey. Also, emerging contaminants in the form of metals is of concern in Brazil (Bernardino et al., 2016). Interestingly, *Typha domingensis* in a constructed wetland of Brazil was found to remove Hg, an emerging contaminant of particular human health concern (Gomes et al., 2014). Tannery wastewater (loaded with diverse emerging contaminants) treatment in Portugal was also shown in constructed wetlands (Calheiros et al., 2014).

In a constructed wetland of Ghana, diverse emerging contaminants, primarily heavy metals like Fe, Cu, Zn, Pb, and Hg, were effectively removed with three wetland macrophytes, *Limnocharis flava*, *Thalia geniculata*, and *Typha latifolia*, thus making them suitable for irrigation in agriculture fields of this country (Anning et al., 2013).

Several global case studies of constructed wetlands in tropical and temperate countries are described concisely (Vymazal, 2002, 2005, 2008, 2010; Vymazal and Kröpfelová, 2008) (Tables 4.3 through 4.6). Here, we present the detailed observations made in constructed wetlands in our global village.

U.S. CASE STUDIES

Nutrient Phytoremediation

Case Study 1: Removal of Total Nitrogen and Phosphorus from Lake Apopka, Orlando, Florida

Brief Relevance of the Case Study   Nutrients like nitrate and phosphate alter the aquatic and wetland ecosystem dynamics through eutrophication. This leads to fish kill due to low dissolved oxygen created by the excessive occurrence of algal blooms and their subsequent utilization of dissolved oxygen. Unfortunately, the biodiversity of wetlands is dramatically reduced and aesthetic pleasure becomes negligible. Further, an excessive nitrate level leads to a human disease called blue baby disease or syndrome.

Impact of Pollution   Lake Apopka became eutrophicated with large amounts of algal biomass. Socioeconomic values and aesthetic pleasure abruptly declined.

Utility of Constructed Wetland   An experiment was designed over an area of 150 ha (more than 300 acres). Removal rates of up to 95% were found for total N and total P (Coveney et al., 1994; Horne, 2000a).

Case Study 2: Drinking Water Treatment: Nitrate Removal Followed by Groundwater Recharge in Prado Wetlands, California

Brief Relevance of the Case Study   As discussed, excessive nitrate levels in drinking water lead to blue baby disease or syndrome. Above 10 mg/L NO^3-N, very young children are susceptible to this fatal disease, which is characterized by poor oxygen transport in the blood. The disease is due to reduction of the ingested nitrate to nitrite in the infant's acid gut. The nitrite then binds with hemoglobin in the bloodstream. Small infants lack the enzymes necessary to reverse the reaction.

TABLE 4.3

Global Context (Examples of Temperate and Tropical Countries) of Environmental Contaminants from Diverse Pollution Sources, Sectors, and Effluents in Treatment Wetlands of FWS Type

S. No.	Source of Contaminants	Global Countries or Sites of Constructed Wetland Projects	Wetland Designers and Researchers
I	Industrial Explosives Food processing Wood waste leachate Sugar factory Refinery Pulp and paper	United States Greece Canada Kenya China and United States China, United States, and Canada	Best et al. (2000) Kapellakis et al. (2004) Masbaugh et al. (2005) Tonderski et al. (2004) Dong and Lin (1994), Gillepsie et al. (2000) Hattano et al. (1993), Xianfa and Chuncai (1994), Goulet and Serodes (2000) Vymazal (2008)
II	Mine drainage Acid coal mine Copper mine Uranium mine Soil heaps	United States Spain Canada Australia United Kingdom	Brodie et al. (1988) De Matos and da Gama (2004) Sobolewski (1996) Overall and Parry (2004) Batty et al. (2005), Vymazal (2008)
III	Municipal or domestic	On global level/ worldwide	Kadlec and Knight (1996), Herouvim et al. (2011), Vymazal et al. (1998), Amaral et al. (2013)
IV	Landfill leachate with emerging contaminants	Sweden Norway United States	Benyamine et al. (2004) Maehlum et al. (1994) Martin et al. (1993)
V	Emerging contaminants/ trace metals in *Phragmites australis*, *Typha australis*, and *Scripus maritimus*-based wetland in southwest Iran	Iran (origin country of Ramsar convention)	Alhashemi et al. (2011)
VI	Storm water runoff Dairy pasture Residential area Airport de-icing Roadside runoff	New Zealand Australia Sweden United Kingdom	Tanner et al. (2005) Bavor et al. (2001) Thoren et al. (2004) Pontier et al. (2004), Vymazal (2008)
VII	FWS treatment wetland	China	Zheng et al. (2016)
VIII	Engineered wetland system (EWS) of SSF or FWS type	Tanzania	Mbuligwe et al. (2011)

(Continued)

TABLE 4.3 (CONTINUED)

Global Context (Examples of Temperate and Tropical Countries) of Environmental Contaminants from Diverse Pollution Sources, Sectors, and Effluents in Treatment Wetlands of FWS Type

S. No.	Source of Contaminants	Global Countries or Sites of Constructed Wetland Projects	Wetland Designers and Researchers
IX	Floating vetiver system; palm oil mill secondary effluent	Malaysia	Darajeh et al. (2016)
X	Treatment wetlands (5 in number; consisting of manure, dairy wastewater, paper pulp, and landfill leachate)	Costa Rica	Nahlik and Mitsch (2006)
XI	Pathogenic microbe elimination using *Typha angustifolia*–based FWS		Khatiwada and Polprasert (1999)

TABLE 4.4

Global Context (Examples of Temperate and Tropical Countries) of Environmental Contaminants from Diverse Pollution Sources, Sectors, and Effluents in Treatment Wetlands of VF Type

S. No.	Source of Contaminants	Global Countries or Sites of Constructed Wetland Projects	Wetland Designers and Researchers
I	Organic contaminants	Portugal	Novais-Martins-Dias (2003)
		Germany	Machte et al. (1997)
		France	Cottin and Merlin (2006)
			Vymazal (2008)
II	Xenobiotics/herbicides	United Kingdom	McKinlay and Kasperek (1999)
III	Municipal or domestic	On global level/ worldwide	Cooper et al. (1996), Vymazal et al. (1998), Kadlec et al. (2000), Vymazal (2008)
IV	Refinery effluent	Pakistan	Aslam et al. (2000)
V	Dairy wastewater	The Netherlands	Veenstra (1998)
VI	Airport runoff	Canada	McGill et al. (2000)
VII	Landfill leachate with emerging contaminants	Australia Germany	Headley et al. (2004), Lindenblatt (2005), Vymazal (2008)
VIII	Refinery effluent with emerging contaminants, e.g., BTEX (benzene, toluene, ethylbenzene, xylene); petroleum hydrocarbons with efficiency above 80%	Temperate or cold climate: Casper, Wyoming	Wallace and Kadlec (2005)

TABLE 4.5
Global Context (Examples of Temperate and Tropical Countries) of Environmental Contaminants from Diverse Pollution Sources, Sectors, and Effluents in Treatment Wetlands of HF Type

S. No.	Source of Contaminants	Global Countries or Sites of Constructed Wetland Projects	Wetland Designers and Researchers
I	Industrial	United States	Thut (1993)
	Pulp and paper	Kenya	Arbira et al. (2005)
	Acid coal mine effluent	Germany	Gerth et al. (2005)
	Chemical	United States	Pantano et al. (2000)
	Lignite pyrolysis	United Kingdom	Sands et al. (2000)
	Distillery or winery	Portugal	Dias et al. (2000)
	Textile	Germany	Wiessner et al. (1999)
	Tannery	India	Billore et al. (2001)
		Italy	Masi et al. (2002)
		South Africa	Sheridan et al. (2006)
		Slovenia	Bulc et al. (2006)
		Australia	Davies and Cottingham (1992)
		Portugal	Calheiros et al. (2007)
		Turkey	Kucuk et al. (2003)
II	Landfill leachate with	Canada	Birckbek et al. (1990)
	emerging contaminants	Norway	Maehlum et al. (1999)
		Slovenia	Bulc (2006)
		United Kingdom	Robinson et al. (1999)
		Poland	Obarska-Pempkowiak et al. (2005)
III	Municipal or domestic	On global level/worldwide	Kadlec and Knight (1996),
	Linear alkyl benzene	United Kingdom	Vymazal et al. (1998)
	sulfonate (LAS)	India	Cooper et al. (1996)
	Pharmaceuticals	Italy	Billore et al. (2002)
		Spain	Del Bubba et al. (2000)
			Matamaros et al. (2005)
IV	Agriculture discharge,		Benyamine et al. (2004)
	runoff, effluent	Australia	Maehlum et al. (1994)
	Pig farm wastewater	China	Martin et al. (1993)
	Fish aquaculture/fish	United Kingdom	Finlayson et al. (1987, 1990)
	farm effluent	Thailand	Wang et al. (1994)
		Taiwan	Gray et al. (1990)
		United States	Kantawanichkul and Neamkam (2003)
		Canada	Lee et al. (2004)
		Germany	Zachritz and Jacquez (1993)
			Comeau et al. (2001)
			Schulz et al. (2003)

(Continued)

TABLE 4.5 (CONTINUED)
Global Context (Examples of Temperate and Tropical Countries) of Environmental Contaminants from Diverse Pollution Sources, Sectors, and Effluents in Treatment Wetlands of HF Type

S. No.	Source of Contaminants	Global Countries or Sites of Constructed Wetland Projects	Wetland Designers and Researchers
V	Dairy/cheese wastewater	Italy	Mantovi et al. (2002, 2003)
		Germany	Kern and Brettar (2002)
		United States	Hill et al. (2003)
		New Zealand	Tanner (1992)
		Denmark	Schierup et al. (1990)
			Vymazal (2008)
VI	HF treatment wetlands	China	Zheng et al. (2016)
VII	Tannery effluent HF treatment wetlands/ constructed wetlands	Portugal	Calheiros et al. (2014)
VIII	Hydrocarbon phytoremediation	Malaysia	Al-Baldawi et al. (2017)
IX	Effluents from domestic uses contaminated with emerging contaminants, specifically metals	Flanders, Belgium	Lesage et al. (2007)

Impact of Pollution   A drinking water supply with excessive nitrate can affect human health, and children are more prone to blue baby disease or syndrome.

Utility of Constructed Wetland   Prado nitrate removal wetlands occupy an area of 200 ha. The Orange County Water District (OCWD) receives its water from Santa Ana River, which was polluted with nitrate. The essence of this form of phytoremediation is the removal of nitrate by conversion to nitrogen gas by bacterial dentrification. Solar-driven phytoremediation in the Prado wetlands acts by producing organic carbon in emergent plants, such as emergent cattails; submergent plants, such as pondweeds; and floating plants, such as duckweed (Horne, 2000a).

Case Study 3: National Park Protection: Removal of
Phosphorus to Prevent Eutrophication in the Everglades
Brief Relevance of the Case Study   The occurrence of algal blooms may lead to the death of wetland flora and fauna or fauna of the adjoining national park.

Impact of Pollution   As discussed in the previous case study, eutrophication perturbs the wetland ecosystem and low dissolved oxygen created by excessive

TABLE 4.6

Global Context (Examples of Temperate and Tropical Countries) of Environmental Contaminants from Diverse Pollution Sources, Sectors, and Effluents in Treatment Wetlands of Hybrid Type (P-Pond)

S. No./Constructed Wetland Type	Source of Contaminants	Global Countries or Sites of Constructed Wetland Projects	Wetland Designers and Researchers
I/FWS-HF	Industrial	China	Wang et al. (1994)
II/VF-HF	Dairy/cheese wastewater	France	Reeb and Werckmann (2005)
III/VF-HF	Municipal or domestic/ sewage wastewater	United States	House and Broome (2000)
		United Kingdom	Burka and Lawrence (1990)
IV/HF-VF	Municipal or domestic/ sewage wastewater	Mexico	Belmont et al. (2004)
V/VF-HF	Landfill leachate with emerging contaminants	Slovenia	Bulc (2000)
VI/VF-HF	Airport runoff	Canada	McGill et al. (2000)
VII/VF-HF	Pig farm wastewater	Thailand	Kantawanichkul and Neamkam (2003)
VIiI/HF-VF	Municipal or domestic/ sewage wastewater	Poland Denmark	Obarska-Pempkowiak et al. (2005) Brix et al. (2003)
VII/VF-HF-FWS-P	Municipal or domestic/ sewage wastewater	Estonia	Mander et al. (2003)
VII/VF-HF-FWS-P	Winery wastewater	Italy	Masi et al. (2003)
IX/HF-VF-VF	Municipal or domestic/ sewage wastewater	Poland	Tuszynska and Obarska-Pempkowiak (2006)
X/HF-FWS	Municipal or domestic/ sewage wastewater	Kenya	Nyakang'o and van Bruggen (1999)
XI/HF-FWS	Landfill leachate with emerging contaminants	Norway	Maehlum et al. (1999)
XIII/FWS-HF	Fish aquaculture	Taiwan	Lin et al. (2002)
XIII/FWS and HF separately and in isolation	FWS and HF treatment wetlands	China	Zheng et al. (2016)
XIV	Vegetated submerged bed constructed wetland (VSBCW)	Nigeria (based on *Phragmites karka* and *Veteveria nigritana*)	Badejo et al. (2015)

occurrence of algal blooms may lead to the death of wetland flora and fauna or fauna of the adjoining national park.

The Florida Everglades National Park faced the problem of receiving nutrient-enriched or -eutrophicated water. Phosphorus removal wetlands of about 17,000 ha intercept and remove total phosphorus. To this end, bladderwort (*Eutricularia*) and blue-green algae proved useful (Horne, 2000a).

Emerging Contaminants: Metals or Selenium

Case Study 4

Brief Relevance of the Case Study   The United States harbors numerous abandoned metal mines (Eger et al., 1993) and coal mines (Brodie, 1993), which result in acid mine drainage. Mediterranean regions have a distinct climate regime composed of cool, wet winters and hot, dry summers (Tsang, 2015). Leached metals from abandoned metal mines into the aquatic ecosystem and wetland perturb the aquatic and wetland ecosystem, specifically fauna health, resulting in death. The reduction in pH due to acid mine drainage tends to make the metals frequently bioavailable, and their impacts are most pronounced. Most of these mines were sulfide ore mines that yielded emerging contaminants like copper, zinc, cadmium, lead, and mercury. Se is a metalloid with properties of both heavy metals and nonmetals, such as sulfur, its close neighbor in the periodic table. It is a vital cofactor in antioxidant removal in mammals, but becomes toxic when more concentrated.

Impact of Pollution: Human Health Impact

Utility of Constructed Wetland   Pertaining to emerging contaminants like heavy metals, constructed wetlands are the only effective and feasible option for eco-sustainable remediation. Interestingly, most heavy metals are immobilized in the anoxic soil of constructed wetlands. *Phragmites, Scirpus, Eichhornia crassipes*, and members of the duckweed family are extremely useful in constructed wetlands for the phytoremediation of such emerging contaminants.

Case Study 5: Phytoremediation of Selenium in Kesterson Marsh, California

Brief Relevance of the Case Study   Se from agriculture, oil and coal industries, photocopying, and automanufacturing has become a major concern in several areas of North America (Science, 1986; Frankenberger and Benson, 1994), particularly California, Colorado, and the Dakotas, and in parts of China. Coal-fired power plants use fuels high in Se, which may pose serious threats to the environment and human health.

Impact of Pollution: Duck Death   The bioconcentration of Se in the food chain by algae resulted in the deaths of birds and concomitantly affected mammals and wildlife.

Utility of Constructed Wetland   Wetland plants immobilized Se, for example, by using a mixture of the giant algae *Chara*, its associated aufwuchs (i.e., the entire attached microbial community of bacteria, algae, fungi, protozoans, and rotifers), and some contribution from emergent wetland plants, such as cattails and alkali bulrush (*Scirpus*). To this end, Se tends to become immobilized and biologically unavailable under anoxic conditions, like metal contaminants (Horne, 2000b).

Emerging Contaminants: Organic Pollutants
Case Study 6: Prado Wetlands, California
Impact of Pollution   All the health impacts discussed for and attributed to organic contaminants apply.

Utility of Constructed Wetland   Wetlands usually result in the remediation of halogenated organics.

Emerging Contaminants: Microbial Pathogens (Horne, 2000b)
In this, there was not any report of use pertaining to the phytoremediation of microbial pathogens through constructed wetlands. Nevertheless, one recent case study paid attention in this context. To this end, in constructed wetlands, to avoid the limitation of large land use a biohedge water hyacinth wetland is frequently used for the remediation of diverse emerging contaminants. Sand filtration may be a very useful mechanism to eliminate emerging contaminants in general and *pathogenic microbes* in particular (Diaz, 2016).

EUROPEAN CASE STUDIES

European case studies are extremely relevant from a global perspective in view of varying climatic and environmental factors. To this end, we should also keep in mind that environmental factors and setup may affect the performance of constructed wetlands in remediating nutrients, organic load, and several emerging contaminants, like PAHs, absorbable organic halogens (AOXs), polychlorinated biphenyls (PCBs), and heavy metals. In view of this fact, it is imperative to consider and discuss the case studies from temperate countries, along with those from tropical countries.

Northern Poland and Southern Sweden
Brief Relevance of the Case Study
Constructed wetlands provide an efficient method for polishing and remediating the landfill leachate in a cost-effective, sustainable way and have potential to remove not only organic carbon and nitrogen compounds, but emerging contaminants, like xenobiotics and heavy metals.

Impact of Pollution
Landfill leachate contains several emerging contaminants, such as PAHs, AOXs, PCBs, heavy metals high in ammonia nitrogen, and chemical oxygen demand (COD) concentrations that pose a serious threat to the environment and human health.

Utility of Constructed Wetland
The constructed wetlands used in these European case studies differed in size, hydraulic regime, type of wetland plants, and type of leachate pretreatment before discharging to the constructed wetlands. The couple of constructed wetlands in

northern Poland consist of two parallel beds with subsurface horizontal flow of leachate (horizontal subsurface flow [HSSF]) planted with reed in one, while the other is a willow plantation that receives leachate after preliminary sedimentation in a retention pond. The best treatment plant was observed in southern Sweden with excellent remediation efficiency (98% total suspended solids, 91% biochemical oxygen demand, 65% COD, and 99.5% N-NH^{4+}), which may be attributed to effective pretreatment of leachate before being discharged into the constructed wetland of a surface flow of leachate (FWS).

One treatment constructed wetland that received wastewater without pretreatment demonstrated less efficiency (27%–61% for biological oxygen demand [BOD] and 2%–35% for COD); however, ammonia nitrogen remediation was satisfactory, with the efficiencies varying from 52% to 89%. Since leachate composition, volume, and quality fluctuations are site specific, the system design should be adapted to these site-specific conditions (Wojciechowska et al., 2010).

RECENT ADVANCES IN WETLAND PLANTS FOR THE REMOVAL OF EMERGING CONTAMINANTS

It has been well established through our previous discussions that constructed wetlands have shown tremendous potential toward emerging contaminants (metal and metalloid removal, volatile organic compounds [VOCs] and hydrocarbons, PAHs, PPCPs, etc.) and pathogenic microbes (Herath and Vithanage, 2015). Further, a global perspective of phytoremediation with wetland plants has been reviewed and represented (Sharma and Pandey, 2014). In view of recent advances, genetic engineering and molecular and omics science are the innovations that provided an impetus to phytotechnologies. Transgenic plants use metal transporters and generate a suite of enzymes to metabolize emerging contaminants through metal-detoxifying chelators, such as MTs and phytochelatins (as elucidated in Figure 2.4). For example, those plants that can synthesize mercuric ion reductase (MerA) and organomercurial lyase (MerB) convert the organo-Hg to metallic Hg to be volatized through the leaf surface (Adams et al., 2013). A type 2 metallotheonein-coding gene (*tyMT*) from *Typha latifolia* (Typhaceae) was transferred to *Arabidopsis thaliana*, and the transgenic *A. thaliana* showed a greater level of tolerance to Cu and Cd (Adams et al., 2013). Glutathione has the ability to prevent cell damage caused by reactive oxygen species (ROS), which is critical in plant defense against a range of abiotic threats; its metabolism is relevant in understanding the tolerance and sequestration of heavy metal ions (Adams et al., 2013). Radial oxygen loss (ROL) from the root to the soil oxygenates the rhizosphere (Adams et al., 2013). Recently, in a constructed wetland of China, *Cyperus alternifolius* interaction with PhACs has been demonstrated through a molecular mechanism (an integrated biochemical and proteomic analysis), and it was noted that ROS could be effectively counteracted by the enhanced antioxidant enzyme activities, and therefore the photosynthetic pigments were ultimately restored (Yan et al., 2016).

Microbial ecology and biotechnology advancement is an integral part of an enhanced phytoremediation pace. To this end, in recent times, the conventional methods of phytoremediation are being augmented using microbial consortia, floating treatment wetlands, constructed wetlands, and hybrid phytoremediation systems (Saeed and Sun, 2012; Vymazal, 2013; Arslan et al., 2014; Arslan, 2016). Nevertheless, the utility of wetland plants in the phytoremediation of emerging contaminants is of paramount importance, and systematic metadata analysis of published literature may assist considerably in a focused phytotechnology approach (Arslan, 2016).

INTERRELATIONSHIP OF WETLAND SYSTEMS AND PLANTS WITH CLIMATE CHANGE AND GREENHOUSE EMISSIONS

It is worth mentioning that in the context of wetland dynamics, wetland systems and plants are inextricably linked with global climate change and greenhouse emissions. Interestingly, the immense biodiversity of wetland plant biomass in treatment wetlands provides an optimum natural environment for the sequestration and long-term storage of carbon dioxide, while wetlands, mangroves, marshes, and swamps are well known for methane (CH_4), a greenhouse gas (Mitsch et al., 2013).

To this end, Mitsch et al. (2013) demonstrated through dynamic modeling of carbon flux the results of 7 detailed studies from temperate and tropical wetlands and 14 other wetland studies from elsewhere and illustrated that on a mere 5%–8% of the terrestrial landscape, wetlands may currently be net carbon sinks of about 830 Tg/year of carbon, mostly in tropical and subtropical countries. In view of this, wetlands should be restored for their immense services, including carbon sequestration, without paying too much attention to the small-scale net radiative sources on the climate due to methane emissions (Mitsch et al., 2013).

Papers of utmost relevance in the context of wetlands were presented during an important event, the Eighth Society of Wetland Scientists (SWS) European Chapter International Conference, "Wetland Systems: Ecology, Functioning and Management," which was held in Padova, Italy, September 1–4, 2013. Discussed were key global issues and findings in geochemistry cycles, plant components, interrelationships between wetlands and climate change, and applicative aspects regarding management and pollution control. The interrelationship with wetland plants and its linkage with greenhouse emissions, bioenergy, and climate change is a present and a future research prospect in the context of Estonia (Mander et al., 2014; Borin and Malagoli, 2015). In combating global climate change, *Cyperus papyrus*, *Cryzopogon zizanoides*, and *Miscanthus* × *giganteus* are wetland plants in subsurface horizontal flow constructed wetlands that are successful in the Mediterranean environment, and to this end, among three macrophytes, *Miscanthus* was the most efficient in carbon sequestration, and hence regulating climate change (Barbera et al., 2014; Borin and Malagoli, 2015).

CONCLUSION

Constructed wetlands have witnessed rapid advancement in the recent past in the remediation of diverse emerging contaminants under varying climatic and environmental factors of tropical and temperate countries. Thus, constructed wetlands are efficient in removing pollutants and emerging contaminants, with the help of wetland plants, in an eco-sustainable way. Further, the use of constructed wetlands offers an environmentally friendly as well as cost-effective approach for the remediation of diverse emerging contaminants. Moreover, plant microbe interaction and the role of rhizospheric organisms or microbial consortia are pertinent to give an impetus to the phytoremediation pace. Also, diverse types of constructed wetlands, depending on environmental and climatic factors, in a global context are being built, and future prospects look bright with the advent of gene manipulation, as well as molecular omics technologies. Further research is required to investigate the interrelationship of wetland systems and climate change.

REFERENCES

Abira MA, van Bruggen JJA, Denny P. 2005. Potential of a tropical subsurface constructed wetland to remove phenol from pre-treated pulp and papermill wastewater. *Wat. Sci. Tech* 51(9):173–175.

Adams A, Raman A, Hodgkins D. 2013. How do the plants used in phytoremediation in constructed wetlands, a sustainable remediation strategy, perform in heavy-metal-contaminated mine sites? *Water Environ J* 27:373–386.

Akinbile CO et al. 2016. Phytoremediation of domestic wastewaters in free water surface constructed wetlands using *Azolla pinnata*. *Int J Phytoremediation* 18(1):54–61.

Al-Baldawi IA et al. 2017. Bioaugmentation for the enhancement of hydrocarbon phytoremediation by rhizobacteria consortium in pilot horizontal subsurface flow constructed wetlands. *Int J Environ Sci Technol* 14:75–84.

Alhashemi AH, Karbassi AR, Kiabi BH, Monavari SM, Nabavi MB. 2011. Accumulation and bioaccessibility of trace element in wetland sediments. *Afr. J. Biotechnol* 10(9):1625–1638.

Amaral R, Ferreira F, Galvao A, and Matos JS. 2013. Constructed wetlands for combined sewer overflow treatment in a Mediterranean country, Portugal. *Water Sci Technol* 67(12):2739–2745.

Anning AK, Korsah PE, Addo-Fordjour P. 2013. Phytoremediation of wastewater with *Limnocharis flava*, *Thalia geniculata* and *Typha latifolia* in constructed wetlands. *Int J Phytoremediation* 15(5):452–464.

Anonymous, 2010. [Internet]. Nature Works. cited 2014 February 18. Available from http://nature-works.net/.

Arias C, Brix H. 2003. Humedales artificiales para el tratamiento de aguas residuales. *Cienc Ing Neogranadina* 13, 17:44.

Armstrong J, Armstrong W. 1990. Light-enhanced convective throughflow increases oxygenation in rhizomes and rhizosphere of *Phragmites australis* (cav) trin ex steud. *New Phytol* 114:121–128.

Armstrong W. 1978. Root aeration in the wetland condition. In *Plant Life in Anaerobic Environments: Processes in Anaerobiosis*, ed. DD Hook, RMM Crawford, 269–298. Ann Arbor, MI: Ann Arbor Science.

Armstrong W, Cousins D, Armstrong J, Turner DW, Beckett PM. 2000. Oxygen distribution in wetland plant roots and permeability barriers to gas-exchange with the rhizosphere: A microelectrode and modelling study with *Phragmites australis*. *Ann Bot* 86:687–703.

Armstrong W, Webb T, Darwent M, Beckett PM. 2009. Measuring and interpreting respiratory critical oxygen pressures in roots. *Ann Bot* 103:281–293.

Aronsson P, Perttu K. 2001. Willow vegetation filters for wastewater treatment and soil remediation combined with biomass production. *For Chron* 77:293–299.

Arslan M. 2016. Importance of metadata in bio/phyto-remediation studies: A way to channel-ize future research through meta-analysis. *Bull Environ Stud* 1(3):88–89.

Arslan M, Afzal M, Amin I, Iqbal S, Khan QM. 2014. Nutrients can enhance the abundance and expression of alkane hydroxylase CYP153 gene in the rhizosphere of ryegrass planted in hydrocarbon-polluted soil. *PLoS One* 9(10):e111208.

Arthur EL, Rice PJ, Anderson TA, Baladi SM, Henderson KLD, Coats JR. 2005. Phytoremediation—An overview. *Crit Rev Plant Sci* 24:109–122.

Aslam MM, Malik M, Baig MA, Qazi IA, Iqbal J. 2007. Treatment performance of compost-based and gravel-based vertical flow wetlands operated identically for refinery waste-water treatment in Pakistan. *Ecol. Eng* 30:34–42.

Badejo A, Sridhar MKC, Coker AO, Ndambuki JM, Kupolati WK. 2015. Phytoremediation of water using *Phragmites karka* and *Veteveria nigritana* in constructed wetland. *Int J Phytoremediation* 17(9):847–852.

Baker AJM. 1981. Accumulators and excluders—Strategies in the response plants to heavy metals. *J Plant Nutr* 3:643–654.

Baker AJM, Brooks RR. 1989. Terrestrial higher plants which hyperaccumulate metallic elements. A review of their distribution, ecology and phytochemistry. *Biorecovery* 1:81–126.

Barbera AC, Borin M, Cirelli GL, Toscano A, Maucieri C. 2014. Comparison of carbon bal-ance in Mediterranean pilot constructed wetlands vegetated with different C4 plant species. *Environ Sci Pollut Res*. doi: 10.1007/s11356-014-2870-3.

Batty LC et al. 2005. Assessment of the ecological potential of mine-water treatment wetlands using a baseline survey of macroinvertebrate communities. *Environ. Poll* 138(3):412–419.

Bavor HJ et al. 2001. Stormwater treatment: Do constructed wetlands yield improved pol-lutant management performance over a detention pond system? *Water Sci. Technol* 44(11–12):565–570.

Belmont MA, Cantellano E, Thompson S, Williamson M, Sánchez A, Metcalfe CD. 2004. Treatment of domestic wastewater in a pilot-scale natural treatment system in central Mexico. *Ecol. Eng* 23:299–311.

Belmont MA, Metcalfe CD. 2003. Feasibility of using ornamental plants (*Zantedeschia aeth-iopica*) in subsurface flow treatment wetlands to remove nitrogen, chemical oxygen demand and nonylphenol ethoxylate surfactants—A laboratory-scale study. *Ecol Eng* 21:233–247.

Benjamin MM, Leckie JO. 1981. Conceptual-model for metal-ligand-surface interactions during adsorption. *Environ Sci Technol* 15:1050–1057.

Benyamine M et al. 2004. Multi-objective environmental management in constructed wet-lands. *Environ. Monit Assess* 90(1–3):171–185.

Bernardino CAR et al. 2016. State of the art of phytoremediation in Brazil—Review and perspectives. *Water Air Soil Pollut* 227:272.

Best EPH et al. 2000. Explosives removal from ground water at the Volunteer Army Ammunition Plant, TN in a small scale wetland modules. In *Wetlands and Remediation*, eds Means JL, Hinchee RE, 365–373. Columbus, Ohio: Battelle Press.

Bialowiec A, Davies L, Albuquerque A, Randerson PF. 2012. The influence of plants on nitro-gen removal from landfill leachate in discontinuous batch shallow constructed wetland with recirculating subsurface horizontal flow. *Ecol Eng* 40:44–52.

Billore SK, Ram H, Singh N, Thomas R, Nelson RM, Pare B. 2002. Treatment performance evaluation of surfactant removal from domestic wastewater in a tropical horizontal sub-surface constructed wetland. In Proc. 8th Internat. Conf. Wetland Systems for Water Pollution Control, University of Dar-es-Salaam, Tanzania and IWA, pp. 393–399.

Billore SK, Singh N, Ram HK, Sharma JK, Singh VP, Nelson RM, Das P. 2001. Treatment of a molasses based distillery effluent in a constructed wetland in central India. *Wat. Sci. Tech* 44(11–12): 441–448.

Birkbeck AE, Reil D, Hunter R. 1990. Application of natural and engineered wetlands for treatment of low-strength leachate. In *Constructed Wetlands in Water Pollution Control*, eds Cooper PF, Findlater BC, 411–418. Oxford, UK: Pergamon Press.

Borin M, Malagoli M. 2015. Ecology, functioning and management of wetland systems. *Environ Sci Pollut Res* 22:2357–2359.

Borin M, Milani M, Salvato M, Toscano A. 2011. Evaluation of *Phragmites australis* (cav.) trin. Evapotranspiration in northern and southern Italy. *Ecol Eng* 37:721–728.

Brisson J, Chazarenc F. 2009. Maximizing pollutant removal in constructed wetlands: Should we pay more attention to macrophyte species selection? *Sci Total Environ* 407:3923–3930.

Brix H. 1994. Functions of macrophytes in constructed wetlands. *Water Sci Technol* 29:71–78.

Brix H. 1997. Do macrophytes play a role in constructed treatment wetlands? *Water Sci Technol* 35:11–17.

Brix H, Arias CA, Johansen NH. 2003. Experiments in a two-stage constructed wetland system: Nitrification capacity and effects of recycling on nitrogen removal. In *Wetlands-Nutrients, Metals and Mass Cycling*, ed Vymazal J, 237–258. Leiden, the Netherlands: Backhuys Publishers.

Brodie GA. 1993. Stages, aerobic constructed wetlands to treat acid drainage: Case history of Fabius Impoundment #1 and overview of the Tennessee Valley Authority's program. In *Constructed Wetlands for Water Quality Improvement*, ed. GA Moshiri, 157–165. Boca Raton, FL: Lewis Publishers.

Brodie GA et al. 1988. Constructed Wetlands for acid drainage control in the Tennessee Valley. Tennessee Valley Authority, Knoxville, TN.

Bulc TG. 2006. Long term performance of a constructed wetland for landfill leachate treatment. *Ecol. Eng* 26:365–374.

Bulc TG, Ojstrsek A, Vrhovšek D. 2006. The use of constructed wetland for textile wastewater treatment. In Proc. 10th Internat. Conf. Wetland Systems for Water Pollution Control, MAOTDR 2006, Lisbon, Portugal, pp. 1667–1675.

Burka U, Lawrence P. 1990. A new community approach to wastewater treatment with higher water plants, in: *Constructed Wetlands in Water Pollution Control*, eds Cooper PF, Findlater BC, 359–371. Oxford, UK: Pergamon Press.

Calheiros C, Rangel AOSS, Castro PML. 2014. Constructed wetlands for tannery wastewater treatment in Portugal: Ten years of experience. *Int J Phytoremediation* 16(9):859–870.

Calheiros CSC, Rangel AOSS, Castro, PKL. 2007. Constructed wetland systems vegetated with different plants applied to the treatment of tannery wastewater. *Wat. Res* 41:1790–1798.

Cambrolle J, Redondo-Gomez S, Mateos-Naranjo E, Figueroa ME. 2008. Comparison of the role of two *Spartina* species in terms of phytostabilization and bioaccumulation of metals in the estuarine sediment. *Mar Pollut Bull* 56:2037–2042.

Comeau Y, Brisson J, Réville JP, Forget C, Drizo A. 2001. Phosphorus removal from trout farm effluents by constructed wetlands. *Wat. Sci. Tech* 44(11–12):55–60.

Cooper P, Boon A. 1987. The use of phragmites for wastewater treatment by the root zone method: The UK approach. In *Aquatic Plants for Water Treatment and Resource Recovery*, ed. SW Reddy, 153–174. Orlando, FL: Magnolia Publishing.

Cooper PF, Job GD, Green MB, Shutes RBE. 1996. *Reed Beds and Constructed Wetlands for Wastewater Treatment*. Medmenham, Marlow, UK: WRc Publications.

Cottin NC, Merlin G. 2006. Removal of polycyclic aromatic hydrocarbons from experimental columns simulating a vertical flow constructed wetland. In Proceedings of 10th International Conference for Water pollution control. MAOTDR Portugal, Lisbon, p. 695.

Coveney MF, Stites DL, Lowe EP, Battoe LE. 1994. Nutrient removal in the Lake Apopka marsh flow-way demonstration project. *Lake Reserv Manag* 9:66.

Darajeh N et al. 2016. Modeling BOD and COD removal from palm oil mill secondary effluent in floating wetland by *Chrysopogon zizanioides* (L.) using response surface methodology. *J Environ Manag* 181:343–352.

Davies TH, Cottingham PD. 1992. The use of constructed wetlands for treating industrial effluent. In Proc. 3rd Internat. Conf. Wetland Systems in Water Pollution Control, IAWQ and Australian Water and Wastewater Association, Sydney, NSW, Australia, pp. 53.1–53.5.

Del Bubba M, Lepri L, Cincinelli A, Griffini O, Tabani F. 2000. Linear alkylbenzensulfonates (LAS) removal in a pilot submerged horizontal flow constructed wetland. In Proc. 7th Internat. Conf. Wetland Systems for Water Pollution Control, University of Florida, Gainesville, pp. 919–925.

De Matos CF, da Gama CD. 2004. Constructed Wetlands for acid mine drainage of a lignite mine-Design and full scale operation. In Proceedings of 9th International Wetlands Conference for Water pollution control. ASTEE Lyon, France, pp. 377–383.

Demirevska-Kepova K, Simova-Stoilova L, Stoyanova Z, Hölzer R, Feller U. 2004. Biochemical changes in barley plants after excessive supply of copper and manganese. *Environ Exp Bot* 52:253–266.

Deng H, Ye ZH, Wong MH. 2004. Accumulation of lead, zinc, copper and cadmium by 12 wetland plant species thriving in metal-contaminated sites in China. *Environ Pollut* 132:29–40.

De Souza MP, Huang CP, Chee N, Terry N. 1999. Rhizosphere bacteria enhance the accumulation of selenium and mercury in wetland plants. *Planta* 209:259–263.

Dias VN, Silva N, Serra H, Imácio MM, Vaz AS. 2000. Constructed wetlands for wastewater treatment in Portugal: Inventory and performance. In Proc. 7th Internat. Conf. Wetland Systems for Water Pollution Control, University of Florida, pp. 845–857.

Diaz PM. 2016. Constructed wetlands and water hyacinth macrophyte as a tool for wastewater treatment: A review. *J Adv Civ Eng* 2(1):1–12.

Dong K, Lin C. 1994. The purification mechanism of the system of wetlands and oxidation ponds. In Proc. 4th International Conf. Wetland Systems for Water Pollution Control, Guangzhou, China, pp. 230–236.

Dorman L, Castle JW, Rodgers JH. 2009. Performance of a pilot-scale constructed wetland system for treating simulated ash basin water. *Chemosphere* 75:939–947.

Doyle MO, Otte ML. 1997. Organism-induced accumulation of iron, zinc and arsenic in wetland soils. *Environ Pollut* 96:1–11.

Drew MC, Saglio PH, Pradet A. 1985. Larger adenylate energy-charge and ATP/ADP ratios in aerenchymatous roots of *Zea mays* in anaerobic media as a consequence of improved internal oxygen transport. *Planta* 165:51–58.

Dunbabin JS, Bowmer KH. 1992. Potential use of constructed wetlands for treatment of industrial wastewaters containing metals. *Sci Total Environ* 111:151–168.

Dwire KA, Kaufman JB, Baham JE. 2006. Plant species distribution in relation to water-table depth and soil redox potential in montane riparian meadows. *Wetlands* 26:131–146.

Eger P, Melchert G, Antonson D, Wagner J. 1993. The use of wetland treatment to remove trace metals from mine drainage. In *Constructed Wetlands for Water Quality Improvement*, ed. GA Moshiri, 171–178. Boca Raton, FL: Lewis Publishers.

El Hamouri B, Nazih J, Lahjouj J. 2007. Subsurface-horizontal flow constructed wetland for sewage treatment under Moroccan climate conditions. *Desalination* 215:153–158.

Emerson D, Weiss JV, Megonigal JP. 1999. Iron-oxidizing bacteria are associated with ferric hydroxide precipitates (Fe-plaque) on the roots of wetland plants. *Appl Environ Microbiol* 65:2758–2761.

Farooqi IH et al. 2008. Constructed wetland system (CWS) for wastewater treatment. In *Proceedings of Taal 2007: The 12th World Lake Conference*, Jaipur (India) 29th October to 2nd November, 2007. p. 1004.

Finlayson M, Chick A, von Oertzen I, Mitchell D. 1987. Treatment of piggery effluent by an aquatic plant filter. *Biol. Wastes* 19:179–196.

Finlayson M, von Oertzen I, Chick AJ. 1990. Treating poultry abattoir and piggery effluents in gravel trenches. In *Constructed Wetlands in Water Pollution Control*, eds Cooper PF, Findlater BC, 559–562. Oxford, UK: Pergamon Press.

Forni C, Tommasi F. 2015. Duckweed: A tool for ecotoxicology and a candidate for phytoremediation. *Curr Biotechnol* 4:000–000.

Frankenberger WT Jr, Benson S, eds. 1994. *Selenium in the Environment*. New York: Marcel Dekker.

Gadd GM. 2004. Microbial influence on metal mobility and application for bioremediation. *Geoderma* 122:109–119.

Gambrell RP. 1994. Trace and toxic metals in wetlands: A review. *J Environ Qual* 23:883–889.

Gbogbo F, Otoo S. 2015. The concentrations of five heavy metals in components of an economically important urban coastal wetland in Ghana: Public health and phytoremediation implications. *Environ Monit Assess* 187:655.

Geller G. 1997. Horizontal subsurface flow systems in the German speaking countries: Summary of long-term scientific and practical experiences; recommendations. *Water Sci Technol* 35:157–166.

Gerth A, Hebner A, Kiessig G, Küchler A, Zellmer A. 2005. Passive treatment of minewater at the Schlema-Alberoda site. In Book of Abstracts of the Internat. Symp. Wetland Pollutant Dynamics and Control, Ghent University, Belgium, pp. 53–54.

Ghosh M, Singh SP. 2005. A review on phytoremediation of heavy metals and utilization of its byproducts. *Appl Ecol Environ Res* 3:1–18.

Gill LW, Ring P, Higgins MNP, Johnston PM. 2014. Accumulation of heavy metals in a constructed wetland treating road runoff. *Ecol Eng* 70:133–139.

Gillepsie WB Jr, Hawkins WB, Rodgers JH Jr, Cano ML, Dorn PB. 2000. Transfers and transformations of zinc in constructed wetlands: Mitigation of a refinery effluent. *Ecol. Eng.* 14:279–292.

Gomes MVT et al. 2014. Phytoremediation of water contaminated with mercury using *Typha domingensis* in constructed wetland. *Chemosphere* 228–233.

Goulet RR, Pick FR. 2001. The effects of cattails (*Typha latifolia* L.) on concentrations and partitioning of metals in surficial sediments of surface-flow constructed wetlands. *Water Air Soil Pollut* 132:275–291.

Goulet R, Sérodes J. 2000. Principles and actual efficiency of constructed wetlands, Field trip guide during 2000 INTECOL Wetland Conference, Québec, Canada.

Gray KR, Biddlestone AJ, Job G, Galanos E. 1990. The use of reed beds for the treatment of agricultural effluents. In *Constructed Wetlands in Water Pollution Control*, eds Cooper PF, Findlater BC, 333–346, Oxford: Pergamon Press.

Green M, Shaul N, Beliavski M, Sabbah I, Ghattas B, Tarre S. 2006. Minimizing land requirement and evaporation in small wastewater treatment systems. *Ecol Eng* 26:266–271.

Gregersen P, Brix H. 2001. Zero-discharge of nutrients and water in a willow dominated constructed wetland. *Water Sci Technol* 44:407–412.

Guittonny-Philippe A et al. 2015. Selection of wild macrophytes for use in constructed wetlands for phytoremediation of contaminant mixtures. *J Environ Manag* 147:108–123.

Hänsch R, Mendel RR. 2009. Physiological functions of mineral micronutrients (Cu, Zn, Mn, Fe, Mo, B, Ci). *Curr Opin Plant Biol* 12:259–266.

Hammer DA. 1994. Guidelines for design, construction and operation of constructed wetland for livestock wastewater treatment. In Proceedings of a Workshop on Constructed Wetlands for Animal Waste Management. Lafayette, IN, pp. 155–181.

Hatano K, Trettin CC, House CH, Wolumn AG. 1993. Microbial populations and decomposition activity in three subsurface flow constructed wetlands. In: *Constructed Wetlands for Water Quality Improvement*, ed. G.A. Moshiri, 541–547. Boca Raton, FL: CRC Press/Lewis Publishers.

Headley TR. 2004. Removal of nutrients and plant pathogens from nursery runoff using horizontal subsurface-flow constructed wetlands, Dissertation, Southern Cross University, Lismore, NSW, Australia.

Headley TR, Davison L, Huett DO, Muller R. 2012. Evapotranspiration from subsurface horizontal flow wetlands planted with *Phragmites australis* in sub-tropical Australia. *Water Res* 46:345–354.

Herath I, Vithanage M. 2015. Phytoremediation in constructed wetlands. In *Phytoremediation: Management of Environmental Contaminants*, ed. AA Ansari et al. Vol. 2. Cham, Switzerland: Springer.

Herouvim E, Christos AS, Tekerlekopoulou A, Vayenas DV. 2011. Treatment of olive mill wastewater in pilot-scale vertical flow constructed wetlands. *Ecol Eng* 37(2011):931–939.

Hill CM, Duxbury JM, Goehring LD, Peck T. 2003. Designing constructed wetlands to remove phosphorus from barnyard run-off: Seasonal variability in loads and treatment. In *Constructed Wetlands for Wastewater Treatment in Cold Climates*, eds Mander Ü, Jenssen P, 181–196. Southampton, UK: WIT Press.

Holmstrom H. 2000. Geochemical processes in sulphidic mine tailings: Field and laboratory studies performed in northern Sweden at the Laver, Stekenjokk and Kristineberg mine sites. PhD dissertation, Lulea University of Technology, Lulea, Sweden.

Horne AJ. 2000a. Phytoremediation by constructed wetlands. In *Phytoremediation of Contaminated Soil and Water*, ed. N Terry, G Banueolos. Boca Raton, FL: CRC Press.

Horne AJP. 2000b. Phytoremediation by constructed wetlands. In *Phytoremediation of Toxic Metals: Using Plants to Clean Up the Environment*, ed. I Raskin, BD Ensley, 13–39. New York: John Wiley.

House CH, Broome SW. 2000. Vertical flow-horizontal flow constructed wetlands combined treatment system design and performance. In Proc. 7th Internat. Conf. Wetland Systems for Water Pollution Control, University of Florida, Gainesville and IWA, pp. 1025–1033.

Huang J, Reneau R, Hageborn C. 2000. Nitrogen removal in constructed wetlands employed to treat domestic wastewater. *Water Res* 34:2582–2588.

ITRC. 2003. Technical and Regulatory Guidance Document for Constructed Treatment Wetlands. Prepared by The Interstate Technology & Regulatory Council Wetlands Team. p. 118. Available at https://www.itrcweb.org/GuidanceDocuments/WTLND-1.pdf

Jonsson J, Jonsson J, Lovgren L. 2006. Precipitation of secondary Fe(III) minerals from acid mine drainage. *Appl Geochem* 21:437–445.

Kadlec RH, Knight RL. 1996. *Treatment Wetlands*. 893 Boca Raton, FL: Lewis Publishers.

Kadlec RH, Wallace SD, eds. 2008. *Treatment Wetlands*. Boca Raton, FL: CRC Press.

Kadlec RH, Knight RL, Vymazal J, Brix H, Cooper PF, Haberl R. 2000. Constructed Wetlands for Water Pollution Control: Processes, Performance, Design and Operation, IWA Scientific and Technical Report No. 8, London.

Kantawanichkul S, Neamkam P. 2003. Optimum recirculation ratio for nitrogen removal in a combined system: Vertical flow vegetated bed over horizontal flow sand bed. In *Wetlands: Nutrients, Metals and Mass Cycling*, ed Vymazal J, 75–86. Leiden, the Netherlands: Backhuys Publishers.

Kapellakis IE, Tsagarakis KP, Angelakis AN. 2004. Performance of free water surface constructed wetlands for olive mill wastewater treatment. In Proc. 9th International Conf. Wetland Systems for Water Pollution Control, ASTEE 2004, pp. 113–120.

Karim MR, Manshadi FD, Karpiscak MM, Gerba CP. 2004. The persistence and removal of enteric pathogens in constructed wetlands. *Water Res* 38:1831–1837.

Kayser A, Wenger K, Keller A, Attinger W, Felix HR, Gupta SK, Schulin R. 2000. Enhancement of phytoextraction of Zn, Cd, and Cu from calcareous soil: The use of NTA and sulfur amendments. *Environ Sci Technol* 34:1778–1783.

Keffala C, Ghrabi A. 2005. Nitrogen and bacterial removal in constructed wetlands treating domestic waste water. *Desalination* 185:383–389.

Kern J, Brettar I. 2002. Nitrogen turnover in a subsurface constructed wetland receiving dairy farm wastewater. In *Treatment Wetlands for Water Quality Improvement*, ed PriesJ, 15–21. Waterloo, Ontario: CH2M Hill Canada Limited.

Khan AG, Kuek C, Chaudhry TM, Koo CS, Hayes W. 2000. Role of plants, mycorrhizae and phytochelators in heavy metal contaminated land remediation. *Chemosphere* 41:197–207.

Khan Z, Doty S. 2011. Endophyte-assisted phytoremediation. *Curr Top Plant Biol* 12:97–105.

Khatiwada NR, Polprasert C. 1999. Kinetics of fecal coliform removal in constructed wetlands. *Water Sci Technol* 40:109–116.

Kitao M, Lei TT, Koike T. 1997. Comparison of photosynthetic responses to manganese toxicity of deciduous broad-leaved trees in northern Japan. *Environ Pollut* 97:113–118.

Kivaisi AK. 2001. The potential for constructed wetlands for wastewater treatment and reuse in developing countries: A review. *Ecol Eng* 16:545–560.

Knight B, Zhao FJ, McGrath SP, Shen ZG. 1997. Zinc and cadmium uptake by the hyperaccumulator *Thlaspi caerulescens* in contaminated soils and its effects on the concentration and chemical speciation of metals in soil solution. *Plant Soil* 197:71–78.

Knight RL, Kadlec RH, Ohlendorf HM. 1999. The use of treatment wetlands for petroleum industry effluents. *Environ Sci Technol* 33:973–980.

Küçük OS, Sengul F, Kapdan IK. 2003. Removal of ammonia from tannery effluents in a reed bed constructed wetland. *Wat. Sci. Tech* 48(11-12):179–186.

Kumar JIN, Soni R, Kumar RN, Bhatt I. 2008. Macrophytes in phytoremediation of heavy metal contaminated water and sediments in Periyej Community Reserve, Gujarat, India. *Turk J Fish Aquat Sci* 8:193–200.

Langergraber G, Simunek J. 2005. Modeling variably saturated water flow and multicomponent reactive transport in constructed wetlands. *Vadose Zone J* 4:924–938.

Lasat MM. 2000. Phytoextraction of metals from contaminated soil: A review of plant/soil/metal interaction and assessment of pertinent agronomic issues. *J Hazard Subst Res* 2:1–25.

Lee BH, Scholz M. 2007. What is the role of *Phragmites australis* in experimental constructed wetland filters treating urban runoff? *Ecol Eng* 29:87–95.

Lee CY, Lee CC, Lee FY, Tseng SK, Liao CJ. 2004. Performance of subsurface flow constructed wetland taking pretreated swine effluent under heavy loads. *Bioresour. Technol* 92:173–179.

Lesage E, Rousseau DPL, Meers E, Tack FMG, De Pauw N. 2007. Accumulation of metals in a horizontal subsurface flow constructed wetland treating domestic wastewater in Flanders, Belgium. *Sci Total Environ* 380:102–115.

Levitt, J. 1980. *Responses of Plants to Environmental Stresses*. New York: Academic Press.

Liao SW, Chang WL. 2004. Heavy metal phytoremediation by water hyacinth at constructed wetlands in Taiwan. *J Aquat Plant Manage* 42:60–68.

Lin LY. 1995. Wastewater treatment for inorganics. In *Encyclopedia of Environmental Biology*, 479–484. Vol. 3. New York: Academic Press.

Lin YF, Jing SR, Lee DY, Wang TW. 2002. Nutrient removal from aquaculture wastewater using a constructed wetlands system. *Aquaculture* 209:169–184.

Lindenblatt C. 2005. Planted soil filters with activated pretreatment for composting-place wastewater treatment. In Proc. Workshop Wastewater treatment in Wetlands. Theoretical and Practical Aspects, eds Toczyáowska I, Guzowska G, 87–93. Gdansk, Poland: GdaÉsk University of Technology Printing Office.

Lv T et al. 2016. Phytoremediation of imazalil and tebuconazole by four emergent wetland plant species in hydroponic medium. *Chemosphere* 148:459–466.

Macci C, Peruzzi E, Doni S, Ianelli R, Masciandaro G. 2014. Ornamental plants for micropollutant removal in wetland systems. *Environ Sci Pollut Res.* doi: 10.1007 /s11356-014-2949-x.

Machate T et al. 1997. Degradation of phenanthrene and hydraulic characteristics in a constructed wetland. *Water Res* 31: 544–560.

Maddison M, Mauring T, Remm K, Lesta M, Mander U. 2009. Dynamics of *Typha latifolia* L. Populations in treatment wetlands in Estonia. *Ecol Eng* 35:258–264.

Maehlum T. 1994. Treatment of landfill leachate in on-site lagoons and constructed wetlands. In: *Proc. 4th Internat. Conf. Wetland Systems for Water Pollution Control*, Guangzhou, China, 553–559.

Maehlum T, Stalnecke P. 1999. Removal efficiency of three cold-climate constructed wetlands treating domestic wastewater: Effects of temperature, seasons, loading rates and input concentrations. *Water Sci.Technol* 40(3):273–281.

Maine MA, Sune N, Hadad H, Sanchez G, Bonetto C. 2006. Nutrient and metal removal in a constructed wetland for waste-water treatment from a metallurgic industry. *Ecol Eng* 26:341–347.

Mander U, Maddison M, Soosaar K, Teemusk A, Kanal A, Uri V, Truu J. 2014. The impact of a pulsing groundwater table on greenhouse gas emissions in riparian grey alder stands. *Environ Sci Pollut Res.* doi: 10.1007/s11356-014-3427-1.

Mander Ü, Teiter S, Lõhmus K, Mauring T, Nurk K, Augustin J. 2003. Emission rates of N_2O and CH_4 in riparian alder forest and subsurface flow constructed wetland. In *Wetlands: Nutrients, Metals and Mass Cycling*, ed Vymazal J, 259–279. Leiden, the Netherlands: Backhuys Publishers.

Manning BA, Fendorf SE, Goldberg S. 1998. Surface structures and stability of arsenic(III) on goethite: Spectroscopic evidence for inner-sphere complexes. *Environ Sci Technol* 32:2383–2388.

Mantovi P, Marmiroli M, Maestri E, Tagliavini S, Piccinini S, Marmiroli N. 2003. Application of a horizontal subsurface flow constructed wetland on treatment of dairy parlor wastewater. *Bioresour. Technol* 88:85–94.

Mantovi P, Piccinini S, Marmiroli N, Maestri E. 2002. Treating dairy parlor wastewater using subsurface-flow constructed wetlands. In *Wetlands and Remediation II*, eds Nehring KW, Brauning SE, 205–212. Columbus, Ohio: Battelle Press.

Marchand L, Mench M, Jacob DL, Otte ML. 2010. Metal and metalloid removal in constructed wetlands, with emphasis on the importance of plants and standardized measurements: A review. *Environ Pollut* 158:3447–3461.

Martin CD, Moshiri, GA, Miller CC. 1993. Mitigation of landfill leachate incorporating in-series constructed wetlands of a closed-loop design. In: *Constructed Wetlands for Water Pollution Improvement*, ed. GA Moshiri, 473–476. Boca Raton, FL: CRC Press/ Lewis Publishers

Masbough A, Frankowski K, Hall KJ, Duff SJB. 2005. The effectiveness of constructed wetland for treatment of woodwaste leachate. *Ecol. Eng* 25:552–566.

Masi F, Conte G, Martinuzzi N, Pucci B. 2002. Winery high organic content wastewaters treated by constructed wetlands in Mediterranean climate. In Proc. 8th Internat. Conf. Wetland Systems for Water Pollution Control, University of Dar-es- Salaam, Tanzania and IWA, pp. 274–282.

Masi F, Martinuzzi N. 2007. Constructed wetlands for the Mediterranean countries: Hybrid systems for water reuse and sustainable sanitation. *Desalination* 215:44–55.

Matagi SV, Swai D, Mugabe R. 1998. A review of heavy metal removal mechanisms in wetlands. *Afr J Trop Hydrobiol Fish* 8:23–35.

Matamoros V, García J, Bayona JM. 2005. Elimination of PPCPs in subsurface and surface flow constructed wetlands. In Book of Abstracts of the Internat. Symp. Wetland Pollutant Dynamics and Control, Ghent University, Belgium, pp. 107–108.

Mbuligwe S, Kaseva ME, Kassenga GR. 2011. Applicability of engineered wetland systems for wastewater treatment in Tanzania—A review. *Open Environ Eng J* 4:18–31, 1874–8295.

McCutcheon SC, Jørgensen SE. 2008. Phytoremediation. *Ecol Eng* 2008:2751–2766.

McGill R, Basran D, Flindall R, Pries J. 2000. Vertical-flow constructed wetland for the treatment of glycol-laden stormwater runoff at Lester B. Pearson International Airport. In: *Proc. 7th International Conf. Wetland Systems for Water Pollution Control*, Lake Buena Vista, Florida, University of Florida and IWA, 1080–1081.

McKinlay RG, Kasperek K. 1999. Observations on decontamination of herbicidepolluted water by marsh plant systems. *Wat. Res* 33:505–511.

Mechora S, Stibilj V, Germ M. 2014. Response of duckweed to various concentrations of selenite. *Environ Sci Pollut Res*. doi: 10.1007/s11356-014-3270-4.

Meharg AA, Cairney JW. 2000. Co-evolution of mycorrhizal symbionts and their hosts to metal-contaminated environments. *Adv Ecol Res* 30:69–112.

Mitsch WJ et al. 2013. Wetlands, carbon, and climate change. *Landscape Ecol* 28:583–597.

Mukhopadhyay S, Maiti SK. 2010. Phytoremediation of metal mine waste. *Appl Ecol Environ Res* 8:207–222.

Nahlik A, Mitsch WJ. 2006. Tropical treatment wetlands dominated by free-floating macrophytes for water quality improvement in Costa Rica. *Ecol. Eng* 28(3):246–247.

Neelima P, Reddy KJ. 2004. Interaction of some heavy metals stress on growth and yield in *Solanum melongena* L. *Ecol Environ Conserv* 10:153–159.

Novais J, Martins-Dias S. 2003. Constructed wetlands for industrial wastewater treatment contaminated with nitroaromatic organic compounds and nitrate at very high concentrations. In 1st International Seminar on the use of aquatic macrophytes for wastewater treatment in Constructed wetlands, eds Dias V, Vymazal J, 277–288. Portugal, Lisbon.

Nyakang'o JB, van Bruggen JJA. 1999. Combination of well-functioning constructed wetland with a pleasing landscape design in Nairobi. Kenya. *Wat. Sci. Tech* 40(3):249–256.

Nyquist J, Greger M. 2009. A field study of constructed wetlands for preventing and treating acid mine drainage. *Ecol Eng* 35:630–642.

Obarska-Pempkowiak H, Gajewska H, Wojciechowska E. 2005. Application, design and operation of constructed wetland systems. In Proc. Workshop Wastewater Treatment in Wetlands. Theoretical and Practical Aspects, eds Toczyáowska I, Guzowska G, 119–125. Gdansk, Poland: GdaĚsk University of Technology Printing Office.

Overall RA, Parry DL. 2004. The uptake of uranium by Eleocharis dulcis (Chinese water chestnut) in the Ranger Uranium Mine constructed wetland filter. *Environ. Poll* 132:307–320.

Pantano J, Bullock R, McCarthy D, Sharp T, Stilwell C. 2000. Using wetlands to remove metals from mining impacted groundwater. In *Wetlands and Remediation*, eds Means JL, Hinchee RE, 383–390. Columbus, Ohio: Battelle Press.

Pontier H et al. 2004. Progressive changes in water and sediment quality in a wetland system for control of highway runoff. *Sci. Total Environ* 319:215–224.

Porteous-Moore F, Barac T, Borremans B, Oeyen L, Vangronsveld J, van der Lelie D, Campbell D, Moore ERB. 2006. Endophytic bacterial diversity in polar trees growing on BTEX-contaminated site: The characterisation of isolates with potential to enhance phytoremediation. *Syst Appl Microbiol* 29:539–556.

Raghukumar C. 2008. Marine fungal biotechnology: An ecological perspective. *Fungal Divers* 31:5–19.

Rai PK. 2007a. Phytoremediation of Pb and Ni from industrial effluents using *Lemna minor*: An eco-sustainable approach. *Bull Biosci* 5(1):67–73.

Rai PK. 2007b. Wastewater management through biomass of *Azolla pinnata*: An ecosustainable approach. *Ambio* 36(5):426–428.

Rai PK. 2008a. Phytoremediation of Hg and Cd from industrial effluents using an aquatic free floating macrophyte *Azolla pinnata*. *Int J Phytoremediation* 10(5):430–439.

Rai PK. 2008b. Heavy-metal pollution in aquatic ecosystems and its phytoremediation using wetland plants: An ecosustainable approach. *Int J Phytoremediation* 10(2):133–160.

Rai PK. 2008c. Mercury pollution from chlor-alkali industry in a tropical lake and its bio-magnification in aquatic biota: Link between chemical pollution, biomarkers and human health concern. *Human Ecol Risk Assess Int J* 14:1318–1329.

Rai PK. 2009. Heavy metal phytoremediation from aquatic ecosystems with special reference to macrophytes. *Crit Rev Environ Sci Technol* 39(9):697–753.

Rai PK. 2010a. Microcosm investigation on phytoremediation of Cr using *Azolla pinnata*. *Int J Phytoremediation* 12:96–104.

Rai PK. 2010b. Phytoremediation of heavy metals in a tropical impoundment of industrial region. *Environ Monit Assess* 165:529–537.

Rai PK. 2010c. Seasonal monitoring of heavy metals and physico-chemical characteristics in a lentic ecosystem of sub-tropical industrial region, India. *Environ Monit Assess* 165:407–433.

Rai PK. 2010d. Heavy metal pollution in lentic ecosystem of sub-tropical industrial region and its phytoremediation. *Int J Phytoremediation* 12(3):226–242.

Rai PK. 2011. *Heavy Metal Pollution and its Phytoremediation through Wetland Plants*. New York: Nova Science Publisher.

Rai PK. 2016. Impacts of particulate matter pollution on plants: Implications for environmental biomonitoring. *Ecotoxicol Environ Saf* 129:120–136.

Rai PK, Mishra A, Tripathi BD. 2010. Heavy metals and microbial pollution of River Ganga: A case study on water quality at Varanasi. *Aquat Ecosyst Health Manag* 13(4):352–361.

Rai PK, Singh M. 2016. *Eichhornia crassipes* as a potential phytoremediation agent and an important bioresource for Asia Pacific region. *Environ Skeptics Critics* 5(1):12–19.

Rai PK et al. 2018. A critical review of ferrate(VI)-based remediation of soil and groundwater. *J Environ Res* 160:420–448.

Rascio N, Navari-Izzo F. 2011. Heavy metal hyperaccumulating plants: How and why do they do it? And what makes them so interesting? *Plant Sci* 180:169–181.

Reeb G, Werckmann M. 2005. First performance data on the use of two pilotconstructed wetlands for highly loaded non-domestic sewage. In *Natural and Constructed Wetlands: Nutrients, Metals and Management*, ed Vymazal J, 43–51. Leiden, the Netherlands: Backhuys Publishers.

Robinson H, Harris G, Carville M, Barr M, Last S. 1999. The use of an engineered reed bed system to treat leachate at Monument Hill landfill site, southern England. In: *Constructed Wetlands for the Treatment of Landfill Leachate*, eds. G Mulamoottil, EA McBean and F Rovers, 71–97. Boca Raton, FL: Lewis Publishers/CRC Press.

Rousseau DPL, Lesage E, Story A, Vanrolleghem PA, Pauw ND. 2008. Constructed wetlands for water reclamation. *Desalination* 218:181–189.

Saeed T, Sun G. 2012. A review on nitrogen and organics removal mechanisms in subsurface flow constructed wetlands: Dependency on environmental parameters, operating conditions and supporting media. *J Environ Manag* 112:429–448.

Samecka-Cymerman A, Stepien D, Kempers AJ. 2004. Efficiency in removing pollutants by constructed wetland purification systems in Poland. *J Toxicol Environ Health A* 67(4):265–275.

Sands Z, Gill LS, Rust R. 2000. Effluent treatment reed beds: results after ten years of operation. In *Wetlands and Remediation*, eds Means JF, Hinchee RE, 273–279. Columbus, Ohio: Battelle Press.

Sarma H. 2011. Metal hyperaccumulation in plants: A review focussing on phytoremediation technology. *J Environ Sci Technol* 4:118–138.

Schierup H-H, Brix H, Lorenzen B. 1990. Wastewater treatment in constructed reed beds in Denmark—State of the art. In *Constructed Wetlands in Water Pollution Control*, eds Cooper PF, Findlater BC, 495–504. Oxford: Pergamon Press.

Schulz C, Gelbrecht J, Rennert B. 2003. Treatment of rainbow trout farm effluents in constructed wetland with emergent plants and subsurface horizontal water flow. *Aquaculture* 217:207–221.

Science. 1986. Selenium threat in the west. *Science* 231:111.

Seo DC, Yu K, DeLaune RD. 2008. Comparison of monometal and multimetal adsorption in Mississippi River alluvial wetland sediment: Batch and column experiments. *Chemosphere* 73:1757–1764.

Sharma P, Pandey S. 2014. Status of phytoremediation in world scenario. *Int J Environ Bioremediat Biodegrad* 2(4):178–191.

Sheoran AS, Sheoran V. 2006. Heavy metal removal mechanism of acid mine drainage in wetlands—A critical review. *Miner Eng* 19:105–116.

Sheridan C, Peterson J, Rohwer J. and Burton S. 2006. Engineering design of subsurface flow constructed wetlands for the primary treatment of winery effluent. In Proc. 10th Internat. Conf. Wetland Systems for Water Pollution Control, MAOTDR 2006, Lisbon, Portugal, pp. 1623–1632.

Sheridan C, Peterson J, Rohwer J, Burton S. 2006. Engineering design of subsurface flow constructed wetlands for the primary treatment of winery effluent, in: Proc. 10th Internat. Conf. Wetland Systems for Water Pollution Control, MAOTDR 2006, Lisbon, Portugal, pp. 1623–1632.

Singh MM, Rai PK. 2016. Microcosm investigation of Fe (iron) removal using macrophytes of Ramsar Lake: A phytoremediation approach. *Int J Phytoremediation* 18(12):1231–1236.

Skousen JG, Ziemkiewicz PF. 1995. *Acid Mine Drainage Control and Treatment*. 2nd ed. National Mine Land Reclamation Publication.

Sobolewski A. 1996. Metal species indicate the potential of constructed wetlands for long-term treatment of metal mine drainage. *Ecol. Eng.* 6(4):259–271.

Song Y. 2003. Mechanisms of lead and zinc removal from lead mine drainage in constructed wetland. PhD thesis, University of Missouri, Rolla.

Stadtman ER, Oliver CN. 1991. Metal-catalyzed oxidation of proteins: Physiological consequences. *J Biol Chem* 266:2005–2008.

Stefanakis AI, Tsihrintzis VA. 2012. Effects of loading, resting period, temperature, porous media, vegetation and aeration on performance of a pilot-scale vertical flow constructed wetlands. *Chem Eng J* 181–182(2012):416–430.

Stein OR, Hook PB. 2005. Temperature, plants, and oxygen: How does season affect constructed wetland performance? *J Environ Sci Health A* 40:1331–1342.

Stoltz E, Greger M. 2002. Accumulation properties of As, Cd, Cu, Pb and Zn by four wetland plant species growing on submerged mine tailings. *Environ Exp Bot* 47:271–280.

Stoltz E, Greger M. 2005. Effects of different wetland plant species on fresh unweathered sulphidic mine tailings. *Plant Soil* 276:251–261.

Stottmeister U, Wiessner A, Kuschk P, Kappelmeyer U, Kastner M, Bederski O, Muller RA, Moormann H. 2003. Effects of plants and microorganisms in constructed wetlands for wastewater treatment. *Biotechnol Adv* 22:93–117.

Sung W, Morgan JJ. 1980. Kinetics and product of ferrous iron oxygenation in aqueous systems. *Environ Sci Technol* 14:561–568.

Tanner CC. 2001. Plants as ecosystem engineers in subsurface-flow treatment wetlands. *Water Sci Technol* 44:9–17.

Tanner CC. 1992. Treatment of dairy farm wastewaters in horizontal and up-flow gravel bed constructed wetlands. In Proc. 3rd Internat. Conf. Wetland Systems in Water Pollution Control, IAWQ and Australian Water and Wastewater Association, Sydney, NSW, Australia, pp. 21.1–21.9.

Tencer Y et al. 2009. Establishment of a constructed wetland in extreme dryland. *Environ Sci Pollut Res* 16:862–875.

Thut RN. 1993. Feasibility of treating pulp mill effluent with a constructed wetland. In: *Constructed Wetlands for Water Quality Improvement*, ed. G.A. Moshiri, 441–447. Boca Raton, FL: Lewis Publishers.

Thoren A et al. 2004.Temporal export of nitrogen from a constructed wetland: influence of hydrology and senescing submerged plants. Ecol. Eng 23(4–5):233–249.

Tiwari S, Dixit S, Verma N. 2007. An effective means of biofiltration of heavy metal contaminated water bodies using aquatic weed *Echhornia crassipes*. *Environ Monit Assess* 129:253–256.

Tonderski KS, Grönlund E, Billgren C. 2005. Management of sugar effluent in The Lake Victoria region. In Proc. Workshop Wastewater treatment in Wetlands. Theoretical and Practical Aspects, eds Toczyáowska I, Guzowska G, 177–184., Gdansk, Poland: GdaĚsk University of Technology Printing Office.

Tsang E. 2015. Effectiveness of wastewater treatment for selected contaminants using constructed wetlands in Mediterranean climates. Master's Projects, University of San Francisco.

Tsihrintzis VA, Akratos CS, Gikas GD, Karamouzis D, Angelakis AN. 2007. Performance and cost comparison of a FWS and a VSSF constructed wetland system. *Environ Technol* 28:621–628.

Turker OC, Bocuk H, Yakar A. 2013. The phytoremediation ability of a polyculture constructed wetland to treat boron from mine effluent. *J Hazard Mater* 252–253:132–141.

Tuszynska A, Obarska-Pempkowiak H. 2006. Impact of organic matter quality on effectivness of contaminants removal in hybrid constructed wetlands. In Proc. 10th Internat. Conf. Wetland Systems for Water Pollution Control, MAOTDR Lisbon, Portugal, pp. 721–728.

Ueckert J, Hurek T, Fendrik I, Niemann EG. 1990. Radial gas-diffusion from roots of rice (*Oryza sativa*) and kallar grass (*Leptochloa fusca* L. Kunth), and effects of inoculation with azospirillum-brasilense Cd. *Plant Soil* 122:59–65.

US EPA. July 1993. Subsurface Flow Constructed Wetlands for Wastewater Treatment: A Technology Assessment. EPA832-R-93-001. Office of Water, Washington, DC.

Veenstra S. 1998. The Netherlands. In *Constructed Wetlands for Wastewater Treatment in Europe*, eds. Vymazal J, Brix H, Cooper PF, Green B, Haberl R, 289–314. Leiden, the Netherlands: Backhuys Publishers.

Vesk PA, Nockold CE, Allaway WG. 1999. Metal localization in water hyacinth roots from an urban wetland. *Plant Cell Environ* 22:149–159.

Victor K, Séka Y, Norbert KK, Sanogo TA, Celestin AB. 2016. Phytoremediation of wastewater toxicity using water hyacinth (*Eichhornia crassipes*) and water lettuce (*Pistia stratiotes*). *Int J Phytoremediation* 18(10):949–955.

Vymazal J. 2002. The use of sub-surface constructed wetlands for wastewater treatment in the Czech Republic:10 years experience. *Ecol Eng* 18:633–646.

Vymazal J. 2005. Horizontal sub-surface flow and hybrid constructed wetlands systems for wastewater treatment. *Ecol Eng* 25:478–490.

Vymazal J. 2008. Constructed wetlands for wastewater treatment: A review. In *Proceedings of Taal 2007: The 12th World Lake Conference*, 965–980.

Vymazal J. 2010. Constructed wetlands for wastewater treatment. *Water* 2:530–549.

Vymazal J. 2011a. Plants used in constructed wetlands with horizontal subsurface flow: A review. *Hydrobiologia* 674:133–156.

Vymazal J. 2011b. Constructed wetlands for wastewater treatment: Five decades of experience. *Environ Sci Technol* 45(1):61–69.

Vymazal J. 2013. The use of hybrid constructed wetlands for wastewater treatment with special attention to nitrogen removal: A review of a recent development. *Water Res* 47(14):4795–4811.

Vymazal J, Kröpfelová L. 2008. *Wastewater Treatment in Constructed Wetlands with Horizontal Sub-Surface Flow.* Dordrecht, the Netherlands: Springer.

Vymazal J, Brix H, Cooper PF, Green MB, Haberl R, eds. 1998. *Constructed Wetlands for Wastewater Treatment in Europe.* Leiden, the Netherlands: Backhuys Publishers.

Wang J, Cai X, Chen Y, Yang Y, Liang M, Zhang Y, Wang Z, Li Q, Liao X. 1994. Analysis of the configuration and the treatment effect of constructed wetland wastewater treatment system for different wastewaters in South China. In Proc. 4th Internat. Conf. Wetland Systems for Water Pollution Control, ICWS'94 Secretariat, Guangzhou, P.R. China, pp. 114–120.

Wallace S, Kadlec R. 2005. BTEX degradation in a cold-climate wetland system. *Water Sci Technol* 51(9):165–172.

Weis JS, Weis P. 2004. Metal uptake, transport and release by wetland plants: Implications for phytoremediation and restoration. *Environ Int* 30:685–700.

Weiss JV, Emerson D, Backer SM, Megonigal JP. 2003. Enumeration of Fe(II)-oxidizing and Fe(III)-reducing bacteria in the root zone of wetland plants: Implications for a rhizosphere iron cycle. *Biogeochemistry* 64:77–96.

Wiessner A, Kuschk P, Sottmeister U, Struckmann D, Jank M. 1999. Treating a lignite pyrolysis wastewater in a constructed subsurface flow wetland. *Water Res* 33:1296–1302.

Wetzel RG. 1975. *Limnology.* Philadelphia: Saunders.

Wetzel RG. 1993. Constructed wetlands: Scientific foundations are critical. In *Constructed Wetlands for Water Quality Improvement*, ed. GA Moshiri, 3–8. Boca Raton, FL: Lewis Publishers.

Williams HG, Bialowiec A, Slater F, Randerson PF. 2010. Diurnal cycling of dissolved gas concentrations in willow vegetation filter treating landfill leachate. *Ecol Eng* 36:1680–1685.

Williams JB. 2002. Phytoremediation in wetland ecosystems: Progress, problems, and potential. *Crit. Rev. Plant Sc.* 21(6):607–635.

Wojciechowska E, Gajewska E, Obarska-Pempkowiak H. 2010. Treatment of landfill leachate by constructed wetlands: Three case studies. *Pol J Environ Stud* 19(3):643–650.

Wong YS, Lam EKH, Tam NFY. 1994. Physiological effects of copper treatment and its uptake in *Festuca rubra* cv. Merlin. *Resour Conserv Recycl* 11:311–319.

Wright DJ, Otte ML. 1999. Wetland plant effects on the biogeochemistry of metals beyond the rhizosphere. Biology and environment. *Proc R Irish Acad* 99B(1):3–10.

Wu H, Zhang J, Ngo HH, Guo W, Hu Z, Liang S, Fan J, Liu H. 2015. A review on the sustainability of constructed wetlands for wastewater treatment: Design and operation. *Bioresour Technol* 175:594–601.

Wu S. 1994. Effect of manganese excess on the soybean plant cultivated under various growth conditions. *J Plant Nutr* 17:991–1003.

Xianfa L, Chuncai J. 1994. The constructed wetland systems for water pollution control in north China. In: *Proc. 4th Internat. Conf. Wetland Systems for Water Pollution Control*, Guangzhou, China,121–128.

Xiao-Bin Z et al. 2007. Phytoremediation of urban wastewater by model wetlands with ornamental hydrophytes. *J. Environ. Sci.* 19:902–909.

Yan Q et al. 2016. Insights into the molecular mechanism of the responses for *Cyperus alternifolius* to PhACs stress in constructed wetlands. *Chemosphere* 164:278–289.

Yang X, Feng Y, He Z, Stoffella PJ. 2005. Molecular mechanisms of heavy metal hyperaccumulation and phytoremediation. *J Trace Elem Med Biol* 18:339–353.

Ye ZH, Baker AJM, Wong MH, Willis AJ. 1997a. Zinc, lead and cadmium tolerance, uptake and accumulation in populations of *Typha latifolia* L. *New Phytol.* 136:469–480.

Ye ZH, Baker AJM, Wong MH, Willis AJ. 1997b. Zinc, lead and cadmium tolerance, uptake and accumulation by the common reed, *Phragmites australis* (Cav.) Trin. Ex Steudel. *Ann Bot* 80:363–370.

Ye ZH, Whiting SN, Lin ZQ, Lytle CM, Qian JH, Terry N. 2001. Removal and distribution of iron, manganese, cobalt and nickel within a Pennsylvania constructed wetland treating coal combustion by-product leachate. *J Environ Qual* 30:1464–1473.

Zachritz WH II, Jacquez RB. 1993. Treating intensive aquaculture recycled water with a constructed wetlands filter system. In: *Constructed Wetlands for Water Quality Improvement*, ed. GA Moshiri, 609–614. Boca Raton, FL: CRC Press/Lewis Publishers.

Zayed A, Growthaman S, Terry N. 1998. Phytoaccumulation of trace elements by wetland plants. I. Duckweed. *J Environ Qual* 27:715–721.

Zhang BY, Zheng JS, Sharp RG. 2010. Phytoremediation in engineered wetlands: Mechanisms and applications. *Procedia Environ Sci* 2:1315–1325.

Zhang DQ, Hua T, Gersberg RM, Zhu J, Ng WJ, Tan SK. 2012. Fate of diclofenac in wetland mesocosms planted with *Scirpus validus*. *Ecol Eng* 49:59–64.

Zhang DQ, Hua T, Gersberg RM, Zhu J, Ng WJ, Tan SK. 2013. Carbamazepine and naproxen: Fate in wetland mesocosms planted with *Scirpus validus*. *Chemosphere* 91(1):14–21.

Zhang Y, He L, Chen Z, Zhang W, Wang Q, Qian M, Sheng X. 2011. Characterization of lead-resistant and ACC deaminase-producing endophytic bacteria and their potential in promoting lead accumulation of rape. *J Hazard Mater* 186:1720–1725.

Zhang Y, Liu J, Zhang J, Wang R. 2013. Energy-based evaluation of system sustainability and ecosystem value of a large-scale constructed wetland in north China. *Environ Monit Assess* 185:5595–5609.

Zheng Y et al. 2016. Effects of interspecific competition on the growth of macrophytes and nutrient removal in constructed wetlands: A comparative assessment of free water surface and horizontal subsurface flow systems. *Bioresour Technol* 207:134–141.

5 Methods/Design in Water Pollution Science of Wetland Systems

DESIGN OF CONSTRUCTED WETLANDS FOR REMEDIATION OF EMERGING CONTAMINANTS

Due to the excessive burden on water resources, treated water is reused for agricultural irrigation purposes, as well as soil amendment through sewage sludge; biosolids or biochar to increase global agricultural output is prevalent in almost all countries of the world. Nevertheless, the repercussions of this reuse of partially treated water and solid sludge components can lead to compartmentalization of the environment with emerging contaminants and NPs (Pico et al., 2017).

The last decades have witnessed rapid advancements in the construct of constructed wetlands to suit or maximize the remediation efficiency of emerging contaminants in varying global climates and environmental factors (WaterAid, 2008; Zhang et al., 2010; Tsang, 2015). In the literature, the term· *constructed wetland* is synonymous with *engineered wetlands*, *artificial wetlands*, *man-made wetlands*, *treatment wetlands*, and *reed beds* (Brix, 1994).

There exists diverse constructed wetlands types: free water surface (FWS) constructed wetlands, subsurface flow (SSF) constructed wetlands, and hybrid constructed wetlands, which incorporate surface and subsurface flows. Further, these wetland systems are categorized as horizontal subsurface flow (HSSF) constructed wetlands and vertical subsurface flow (VSSF) constructed wetlands, or a hybrid of the two (Figures 5.1 through 5.4). A designer approach of constructed wetlands is extremely relevant, and it varies with the type and concentrations of the emerging contaminants of particular concern. A designer guide to constructed wetlands is shown in Table 5.1. Further, the benefits and constraints of different constructed wetland designs are explained in Table 5.2.

It is worth mentioning that Dr. Kathe Seidel used wetland plants for the first time in a European country, Germany, in an experimental setup (Seidel, 1964, 1976). In recent times, horizontal flow constructed wetlands have paved the way to vertical flow ones in view of the excessive load of emerging contaminants. The Netherlands was credited with the FWS constructed wetland in 1967, with a total area of 1 ha (Veenstra, 1998). However, technological implementation of the FWS constructed wetland started in North America in the 1960s, and Odum designed coastal lagoons built with cypress to ameliorate sewage wastewater (Odum et al., 1977). The first engineered method for constructed wetlands was developed by the University of Michigan for the Houghton Lake Project. Interestingly, this project was suited to temperate and cold climates (Kadlec and Tilton, 1979; Kadlec et al., 1975). Gradual but concrete

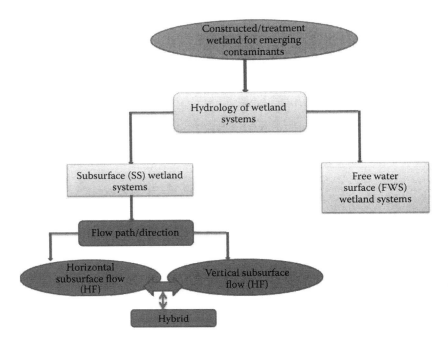

FIGURE 5.1 Diverse design of global engineered constructed wetlands for the removal or remediation of emerging contaminants.

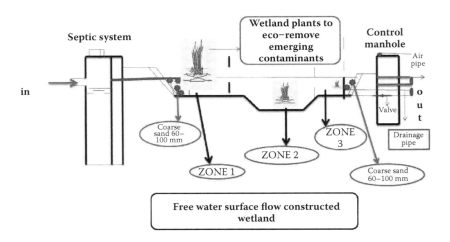

FIGURE 5.2 Design of a typical FWS constructed wetland in the treatment and remediation of emerging contaminants.

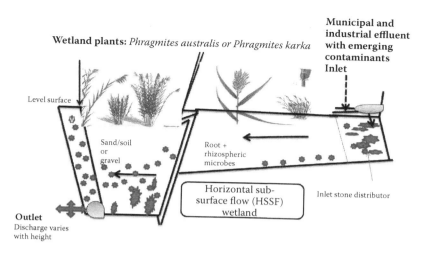

FIGURE 5.3 Design of a typical HSSF constructed wetland in the treatment and remediation of emerging contaminants with *Phragmites australis*, a wetland plant.

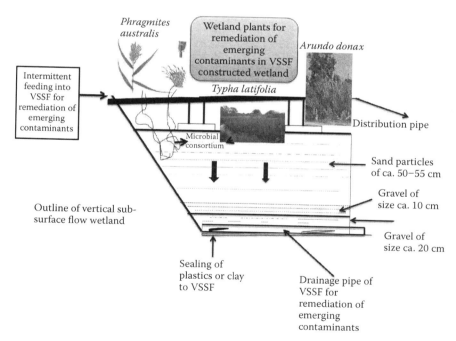

FIGURE 5.4 Design of a typical VSSF constructed wetland in the treatment and remediation of emerging contaminants with *Phragmites australis*, *Typha latifolia*, and *Arundo donax*, wetland plants.

TABLE 5.1
Designer Guide to Engineering Constructed Wetlands

Design modifications	Aeration in/under substrate beds to increase aerobic biodegradation rates
	Use of engineered SSF substrates in place of gravel to adsorb contaminants and control hydraulic loading
Process additions	Chemical and energy addition (e.g., low-grade heat)
	Dilution, alkaline streams
Vegetation changes	Plant harvesting for nutrient removal, phytoremediating plants, stress-resistant species
Advanced operation methods	Recycling of effluents, intermediate streams, separation of competing reactions into different cells

TABLE 5.2
Diverse Designs of Constructed Wetland Systems: Pros and Cons

Constructed Wetland Type	Benefits	Constraints/Limitations
Free water surface flow	• High removal rates of TSS and organic matter and pathogenic microbes • Provides ancillary benefits (wildlife habitat, flood protection) • Aesthetically desirable	• Low removal of nutrients • Land-intensive • Odor can be an issue • Communicable diseases spread or risk due to possible breeding of vectors • Long start-up time
Horizontal subsurface flow	• Low operating costs • High potential for denitrification abilities due to anaerobic conditions • Low odor and mosquito issues • High removal rates of TSS and organic matter	• Low removal of pathogenic microorganisms and nutrients • High risk of clogging • Land-intensive, like FWS flow
Vertical subsurface flow	• Not as prone to clogging as HSSF systems • Low odor and mosquito issues, as applicable to HSSF systems • High potential for nitrification abilities due to aeration • Not land-intensive, as applicable to HSSF and FWS flow systems • High removal rates of TSS and organic matter	• Requires more maintenance than HSSF and FWS systems • Low removal of pathogenic microorganisms and nutrients • May require an electrical energy source for the trickling method • Requires precise measurements of wastewater inputs

Note: TSS, total suspended solids.

progress has been made in the advancement of constructed wetlands through the efforts of Wieder (1989), Kadlec and Knight (1996), Vymazal (2006) and Gunes et al. (2012).

ANALYTICAL OR INSTRUMENTATION TECHNOLOGIES TO ASSESS EMERGING CONTAMINANTS AND CHARACTERIZE NANOPARTICLES AND NANOMATERIALS

Characterization of emerging contaminants in the context of size, structure, composition, coatings, and so forth, is pertinent in view of commercial applications in diverse sectors and the elucidation of eco-toxicity. Analytical tools for emerging contaminants and NPs in the environment and plant tissues are mentioned in detail elsewhere (Pico et al., 2017).

Figures 5.5 and 5.6 highlight the characterization of emerging contaminants based on molecular, separation, microscopy, and spectrometric techniques. In addition to characterization of emerging contaminants, these advanced instrumentation

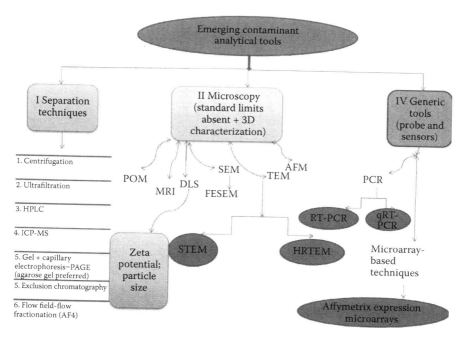

FIGURE 5.5 Emerging contaminant and NP characterization and analysis through separation, microscopy, and spectrometric and molecular tools. HPLC, high-performance liquid chromatography; ICP-MS, inductively coupled plasma mass spectrometry; PAGE, polyacrylamide gel electrophoresis; POM, polarized optical microscopy; MRI, magnetic resonance imaging; DLS, dynamic light scattering; SEM, scanning electron microscopy; FESEM, field emission scanning electron microscopy; TEM, transmittance electron microscopy; HRTEM, high-resolution TEM; STEM, scanning TEM; AFM, atomic force microscopy; PCR, polymerase chain reaction; RT-PCR, real-time PCR; qRT-PCR, quantitative real-time PCR.

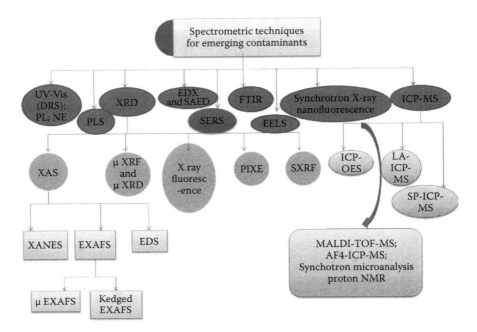

FIGURE 5.6 Advanced spectrometric instrumentation for analysis elucidation and characterization of emerging contaminants and NPs. ICP-MS, inductively coupled plasma mass spectrometry; UV-Vis, ultraviolet-visible; DRS, diffuse reflectance spectrometry; PL, photoluminescence; PLS, photoluminescence spectroscopy; NE, null ellipsometer; XRD, x-ray diffraction; XAS, x-ray absorption spectroscopy; XANES, x-ray absorption spectroscopy near edge structure; EXAFS, extended x-ray absorption spectroscopy fine structure; PIXE, proton-induced x-ray emission; SXRF, synchrotron x-ray fluorescence; EDS, element-dispersive spectrometry; EDX, energy-dispersive x-ray; SAED, selected area electron diffraction; SERS, surface-enhanced Raman spectroscopy; FTIR, Fourier transform infrared; EELS, electron energy loss spectroscopy; ICP-OES, inductively coupled plasma optical emission spectroscopy; LA-ICP-MS, laser ablation ICP-MS; SP-ICP-MS, single-particle ICP-MS; AF4-ICP-MS, flow field-flow fractionation ICP-MS; NMR, nuclear magnetic resonance; MALDI-TOF-MS, matrix-assisted laser desorption/ionization time-of-flight mass spectrometry.

technologies play a vital role in advancing plant–emerging contaminant interactions in the environment and energy sector, which we discuss in different sections. Furthermore, separation NP techniques are initiated with centrifugation and ultrafiltration; however, gradual improvements to address the limitations have led to high-performance liquid chromatography (HPLC) (also used in integration or coupling with inductively coupled plasma mass spectrometry (ICP-MS) for better separation of emerging contaminants and NPs), gel and capillary electrophoresis (including polyacrylamide gel electrophoresis [PAGE]), size exclusion chromatography (SEC), cross-flow ultrafiltration (CFUC), and flow field-flow fractionation (AF4). Also, microscopy techniques like the zeta potentiometer have proved useful in elucidating NP aggregation, size, and shape. Nevertheless, quantitative characterization and time consumption are prime constraints in microscopy. In microscopy, polarized optical microscopy

(POM), atomic force microscopy (AFM), and transmittance electron microscopy (TEM) have proved to be more useful than scanning electron microscopy (SEM). To this end, spectrometric techniques are the most widely used in emerging contaminant characterization, which were initiated with economical UV-vis and switched over to x-ray diffraction (XRD), energy-dispersive x-ray (EDX), Fourier transform infrared (FTIR), ICP-MS (single-particle ICP-MS [SP-ICP-MS] or AF4-ICP-MS), and so forth.

LIMITATIONS OF INVASIVE WETLAND PLANTS IN WETLAND DESIGN: ECO-SUSTAINABLE SOLUTION

Interestingly, certain wetland plants and macrophytes, like *Eichhornia crassipes* (water hyacinth), may be a matter of concern to constructed wetland systems or other aquatic environment health in view of being invasive in nature due to their rapid pace of reproduction, in addition to acting as a worthy candidate or tool in phytoremediation. To this end, a designer plant approach for in situ microbial degradation of emerging contaminants within wetland plants/macrophytes and their subsequent use in bioenergy production may be a clean option that not only boosts the option of acquiring sustainability in the energy sector but also assists in environmentally friendly management of diverse environmental pollutants. Further, by coupling phytoremediation with bioenergy and derivatives, we can enhance the economic returns of this phytotechnology. My past research and articles on invasion ecology have emphasized that declaring a wetland and terrestrial plants as invasive does not mean that they can not be involved in environmental management through phytotechnologies and phytoremediation (Rai, 2012, 2013, 2015a–d, 2016; Rai and Chutia, 2016; Rai and Singh, 2015, 2016).

ANALYTICAL METHODS TO ASSESS THE WATER QUALITY OF WETLANDS

COLLECTION OF SAMPLES

Water

Samplings were usually done in the morning between 6:30 and 9:30 a.m., and the samples were immediately transported to the laboratory and analyzed. Wide-mouth bottles were used to collect samples for the analysis. Tags or labels for each batch and sample were given for easy identification. Analysis of the water was carried out using "standard methods" (APHA, 2005). Thirteen water quality parameters were studied: temperature, pH, transparency, total solids (TS), dissolved oxygen (DO), biological oxygen demand (BOD), acidity, alkalinity, chloride, hardness, turbidity, nitrogen-nitrate, and phosphate, along with the heavy metal analyses. The temperature was measured by a digital thermometer and expressed in degrees Celsius. The pH value was determined by a Hanna digital pH meter. Transparency was measured using a Secchi disc. TS was measured by using filtration and an evaporation method, DO and BOD by using the Winkler titrimetric method, and alkalinity and acidity by using the potentiometric titration method. The chloride content was measured by using Mohr's argentiometric method, total hardness by using the EDTA titration method, turbidity by using a digital turbidity meter, nitrate content by using the phenol di-sulfonic acid method, and phosphate content by using the stannous chloride method.

Macrophytes and Wetland Plants

Macrophyte sampling was also done in the morning between 6:30 and 9:30 a.m., and the samples were immediately transported to the laboratory and analyzed. Transparent polybags were used for macrophyte sampling. The fresh macrophytes were kept inside the polybags with water in order to prevent drying of the fresh macrophyte samples. Fresh macrophyte samples were collected from all sampling sites in different seasons of the year, depending on their growth stages; thoroughly washed with tap water to eliminate the remains of lake sediments; and placed in corresponding lake water that received 8 hours of fluorescent light per day. The fresh macrophyte samples were weighed and kept in an oven for drying, and the temperature was maintained at 80°C for 24 hours. The dried macrophyte samples were again weighed and then crushed into powder. The powdered macrophyte samples were then digested using the di-acid method.

Analysis

Water

Samples were analyzed within 1–6 hours of collection for physical parameters, such as temperature, turbidity, transparency, TS, total suspended solids, and total dissolved solids, and within 24 hours for chemical parameters, such as pH, total hardness, acidity, alkalinity, chloride, nitrate, phosphate, and DO, except for BOD, which took 5 days due to incubation. The methods for the analysis of various physicochemical characteristics are as follows.

1. Temperature (Celsius thermometer)
   The temperature of water is measured by using a small centigrade thermometer with a precision of 0.1°C or by means of a digital thermometer.
2. Turbidity (nephelometer)
   Turbidity is measured either by its effect on the transmission of light, which is called turbidity, or by its effect on the scattering of light, which is called nephelometry. A nephelometer is used for measuring turbidity.
3. Transparency (Secchi disc)
   Transparency is measured by using a Secchi disc. It can be calculated by using the formula

$$SDT = A + B/2$$

   where:
   SDT = Secci disc transparency (cm)
   A = depth at which the Secchi disc disappears (cm)
   B = depth at which the Secchi disc reappears (cm)

4. Total solids
   TS is measured by using filtration and an evaporation method.
   TS (g/L) can be calculated by using the formula

$$TS = W_1 - W_2$$

where:

W_1 = final weight of the crucible
W_2 = initial weight of the crucible
V = volume of the water sample evaporated (mL)

5. Dissolved oxygen

DO is measured by using Winkler's modified azide method. It can be calculated by using the formula

$$DO(mg/L) = \frac{(0.2 \times 1000) \times \text{Volume of thiosulfate}}{100}$$

6. Biological oxygen demand (Winkler's modified azide method)

BOD can be measured by using Winkler's modified azide method. Determination of DO is required for the analysis of BOD.

The DO content can be calculated by using the formula given above. BOD can be calculated by using the formula

$$BOD = DO_1 - DO_2$$

where:

DO_1 = DO taken before incubation of the sample
DO_2 = DO taken after incubation of the sample

7. pH (digital electronic pH meter)

The pH value of the natural water is an important index of acidity and alkalinity. pH can be measured by using the digital pH meter.

8. Hardness (EDTA titration method)

The hardness is measured using the EDTA titration method. It can be calculated by using the following formulas for $CaCO_3$.
   a. Total hardness

$$\text{Total Hardness as } CaCO_3, \text{ mg/L} = \frac{C \times D \times 1000}{\text{Volume of sample taken}}$$

where:

C = volume of EDTA required by sample
D = milligrams of $CaCO_3$ equivalent to 1.0 mL of EDTA titrant

   b. Calcium hardness

$$\text{Calcium hardness as } CaCO_3 = \frac{A' \times D \times 1000}{\text{Volume of sample taken}}$$

where:
  A′ = volume of EDTA used by sample
  D = milligrams of $CaCO_3$ equivalent to 1.0 mL of EDTA titrant

  c. Magnesium hardness

Magnesium hardness as $CaCO_3$ = total hardness as $CaCO_3$ – Ca hardness as $CaCO_3$

9. Acidity (potentiometric titration method)
     Acidity is measured by using the potentiometric titration method.
     It can be calculated by using the formula

$$\text{Total acidity} = \frac{(A+B) \times 1000}{\text{Volume of sample taken}}$$

  where:
     A = acidity due to mineral
     B = acidity due to CO_2

10. Alkalinity (potentiometric titration method)
      Alkalinity is measured by using the potentiometric titration method.
      It can be calculated by using the formula

$$\text{Total alkalinity} = \frac{(A+B) \times 1000}{\text{Volume of sample taken}}$$

  where:
     A = alkalinity due to phenolphthalein
     B = alkalinity due to methyl orange

11. Chloride (Mohr's argentometric method)
      Chloride content can be measured by using Mohr's argentiometric method.
      It can be calculated by using the formula

$$\text{Chloride content (mg/L)} = \frac{\text{Volume of titrant} \times 0.0114 \times 35.45 \times 1000}{\text{Volume of sample taken}}$$

12. Nitrate (phenol di-sulfonic acid method)
      The phenol di-sulfonic acid method is used for the estimation of nitrate N. The steam-dried water samples are dissolved in 2 mL of phenol di-sulfonic acid and further the medium is turned to alkaline by adding 10 mL of ammonium hydroxide. The development of yellow color indicates the presence of nitrate-N. The color intensity is proportional to the amount of

nitrate-N. The optical density is measured with the help of a colorimeter at 410 nm. The final calculations are made with a known standard graph.

$$\text{Nitrogen-N} = \frac{\text{Reading of standard} \times 1000}{\text{mL sample}} \text{ mg/L}$$

$$C_6H_3OH\,(HSO_3)_2 + HNO_3 \rightarrow C_6H_2OH_9HSO_3)_2\,NO_2 + H_2O$$
(Nitrophenol sulfonic acid)

13. Phosphate (stannous chloride method)

The stannous chloride method is used for the determination of phosphate concentration in a water sample. Ammonium molybdate solution (4 mL) and 10 drops of stannous chloride solution are added to the 50 mL sample water and measured to 100 mL with the distilled water. The development of blue color indicates the presence of phosphate-P. The color intensity is proportional to the amount of phosphorus present and is measured in terms of optical density with the help of a colorimeter at 680 nm. The final calculation is made with the help of a standard graph, prepared from a known concentration of phosphate in solution.

$$H_3PO_4 + 12H_2MoO_4 \rightarrow H_3P(MoO_{10})_4 + 12H_2O$$
(Blue)

HEAVY METALS: ANALYTICAL METHODS

Water samples are acidified with sulfuric acid and kept for analysis. Macrophyte and wetland plant samples are digested using Di-acid methods for macrophyte digestion.

For the Di-acid method for macrophyte sample digestion, 0.5–1.0 g of dried and processed macrophyte samples is weighed in a 150 mL conical flask. Add 10 mL of concentrated HNO_3, place a funnel on the flasks, and keep for about 6–8 hours overnight in a covered place for predigestion. After predigestion, when the solid sample is no longer visible, add 10 mL of concentrated HNO_3 and 3 mL of $HClO_4$. The flask is then kept on a hot plate and heated at about 100°C for 1 hour, and then the temperature is raised to about 200°C. The digestion is continued until the contents become colorless and only white dense fumes appear. The acid content is reduced to about 2–3 mL, without letting it be dried up by continued heating at the same temperature. The flasks are then removed from the hot plate and cooled, and about 30 mL of distilled water is added. The content is then filtered through Whatman filter paper no. 42 into a 100 mL volumetric flask. The volume is then made up to 100 mL. Digested samples are analyzed with an atomic absorption spectrophotometer (AAS) and microwave-induced plasma atomic emission spectrophotometer (MP-AES) (Rai, 2007a,b, 2008a–c, 2009, 2010a–d, 2011; Rai et al., 2010; Singh and Rai, 2016).

For the above water quality parameters, the methods described in *Standard Methods for the Examination of Water and Wastewater* by the American Public Health Association (APHA, 2005) have been adopted for their analysis.

PHYTOSOCIOLOGICAL ANALYSIS OF WETLAND VEGETATION, PLANTS, AND MACROPHYTES

Vegetation analysis is carried out by following the standard methods outlined in Misra (1968), Kershaw (1973), and Mueller-Dombois and Ellenberg (1974). Harvest methods are adopted for phytosociological analysis pertaining to the macrophytes, and quadrats (1 × 1 m) are used. Macrophytic diversity is usually calculated using the following indices:

1. Simpson's index of dominance, Cd

$$Cd = \sum_{i=0}^{n} pi^2$$

where pi = proportion of individual in the ith species.

As Simpson's index value decreases, diversity decreases. Simpson's index is therefore usually expressed as $1 - Cd$ (Simpson, 1949).

2. Sorenson's similarity index, β

$$\beta = \frac{2c}{S_1 + S_2}$$

where:
$S1$ = number of species in community 1
$S2$ = number of species in community 2
c = number of species common to both communities (Sorenson, 1948)

3. Shannon–Wiener diversity index, H'

$$H' = -\sum_{5=1}^{s} pi \, lnpi$$

where:
H' = Shannon–Wiener diversity index
Pi = proportion of individuals in the ith species, that is, ni/N (Shannon and Weaver, 1949)

DESIGN OF PHYTOREMEDIATION EXPERIMENTS USING MACROPHYTES

Phytoremediation experiments with four selected macrophytes are conducted for contamination analysis. The macrophytes may be *Eichhornia crassipes*, *Lemna minor*, *Pistia stratiotes*, and *Salvinia cucullata* (in our case, explained in Chapter 6). Macrophytes are selected according to there being adequate availability at all the sites and their active role in phytoremediation, as discussed in the review of the literature.

PHYTOREMEDIATION OF IRON (FE): DESIGN

As all the metals except Fe were well below the permissible limit prescribed by the World Health Organization (WHO) (see Table 6 in the appendix), I planned a microcosm investigation of the phytoremediation of Fe only in view of the current context of metal pollution in Loktak Lake (details are given in Chapter 6 and 7).

Macrophytes, that is, *Eichhornia crassipes*, *Lemna minor*, *Pistia stratiotes*, and *Salvinia cucullata*, are separately kept in 39 numbers of 5 L glass aquarium especially constricted for the microcosm experiments. Fifty grams of the fresh weight of each macrophyte sample was used in 1 L of each Fe aqueous solution. $FeSO_4$ was taken as the Fe concentration for the experiment. (Phytoremediation experiments are discussed in Chapter 7.) After the calculation, $FeSO_4$ is measured keeping in mind that the Fe content in the $FeSO_4$ should be maintained at 1, 3, and 5 mg/L, respectively, in all aquariums in triplicate. For each set of experiments, 1 L each of the 1, 3, and 5 mg/L solutions of Fe metals was poured into each aquarium, respectively. Further, each of these solutions was treated with the respective macrophyte for 4, 8, and 12 days. Each set of experiments was repeated thrice, with a control for all the concentrations where the solution was not treated with a plant. The water samples were collected and analyzed after the experiment. The average data of the three replications were taken into consideration.

REFERENCES

APHA (American Public Health Association). 2005. *Standard Methods for the Examination of Water and Wastewater*. 21st ed. Washington, DC: APHA, American Water Works Association, and Water Environment Federation.

Brix H. 1994. Use of constructed wetlands in water pollution control: Historical development, present status, and future perspectives. *Water Sci Technol* 30(8):209–223.

Gunes K, Tunciper B, Ayaz S, Drizo A. 2012. The ability of free water surface constructed wetland system to treat high strength domestic wastewater: A case study for the Mediterranean. *Ecol Eng* 44:278–284.

Kadlec RH, Knight RL. 1996. *Treatment Wetlands*. Boca Raton, FL: Lewis Publishers.

Kadlec RH, Richardson CJ, Kadlec JA. 1975. The effects of sewage effluent on wetland ecosystems. Semiannual Report No. 4 (NTIS PB 2429192). Ann Arbor, MI: University of Michigan.

Kadlec RH, Tilton DL. 1979. The use of freshwater wetlands as a tertiary wastewater treatment alternative. *CRC Crit Rev Environ Control* 9:185–212.

Kershaw KA. 1973. *Quantitative and Dynamic Plant Ecology*. London: Edward Arnold Ltd.

Misra R. 1968. *Ecology Work Book*. Calcutta: Oxford Publishing Company.

Mueller-Dombois D, Ellenberg H. 1974. *Aims and Methods of Vegetation Ecology*. Hoboken, NJ: John Wiley & Sons.

Odum HT, Ewel KC, Mitsch WJ, Ordway JW. 1977. Recycling treated sewage through cypress wetlands in Florida. In *Wastewater Renovation and Reuse*, ed. FM D'Itri, 35–67. New York: Marcel Dekker.

Pico Y et al. 2017. Analysis of emerging contaminants and nanomaterials in plant materials following uptake from soils. *Trends Analyt Chem* 94:173–189.

Rai PK. 2007a. Phytoremediation of Pb and Ni from industrial effluents using *Lemna minor*: An eco-sustainable approach. *Bull Biosci* 5(1):67–73.

Rai PK. 2007b. Wastewater management through biomass of *Azolla pinnata*: An ecosustainable approach. *Ambio* 36(5):426–428.

Rai PK. 2008a. Phytoremediation of Hg and Cd from industrial effluents using an aquatic free floating macrophyte *Azolla pinnata*. *Int J Phytoremediation* 10(5):430–439.

Rai PK. 2008b. Heavy-metal pollution in aquatic ecosystems and its phytoremediation using wetland plants: An ecosustainable approach. *Int J Phytoremediation* 10(2):133–160.

Rai PK. 2008c. Mercury pollution from chlor-alkali industry in a tropical lake and its bio-magnification in aquatic biota: Link between chemical pollution, biomarkers and human health concern. *Hum Ecol Risk Assess Int J* 14:1318–1329.

Rai PK. 2009. Heavy metal phytoremediation from aquatic ecosystems with special reference to macrophytes. *Crit Rev Environ Sci Technol* 39(9):697–753.

Rai PK. 2010a. Microcosm investigation on phytoremediation of Cr using *Azolla pinnata*. *Int J Phytoremediation* 12:96–104.

Rai PK. 2010b. Phytoremediation of heavy metals in a tropical impoundment of industrial region. *Environ Monit Assess* 165:529–537.

Rai PK. 2010c. Seasonal monitoring of heavy metals and physico-chemical characteristics in a lentic ecosystem of sub-tropical industrial region, India. *Environ Monit Assess* 165:407–433.

Rai PK. 2010d. Heavy metal pollution in lentic ecosystem of sub-tropical industrial region and its phytoremediation. *Int J Phytoremediation* 12(3):226–242.

Rai PK. 2011. *Heavy Metal Pollution and Its Phytoremediation through Wetland Plants*. New York: Nova Science Publishers.

Rai PK. 2012. Assessment of multifaceted environmental issues and model development of an Indo-Burma hot spot region. *Environ Monit Assess* 184:113–131.

Rai PK. 2013. *Plant Invasion Ecology: Impacts and Sustainable Management*. New York: Nova Science Publishers.

Rai PK. 2015a. Paradigm of plant invasion: Multifaceted review on sustainable management. *Environ Monit Assess* 187:759–785.

Rai PK. 2015b. What makes the plant invasion possible? Paradigm of invasion mechanisms, theories and attributes. *Environ Skeptics Critics* 4(2):36–66.

Rai PK. 2015c. Concept of plant invasion ecology as prime factor for biodiversity crisis: Introductory review. *Int Res J Environ Sci* 4(5):85–90.

Rai PK. 2015d. *Environmental Issues and Sustainable Development of North East India*. Saarbrücken, Germany: Lambert Academic Publisher.

Rai PK. 2016. Biodiversity of roadside plants and their response to air pollution in an Indo-Burma hotspot region: Implications for urban ecosystem restoration. *J Asia Pac Biodivers* 9:47–55.

Rai PK, Chutia B. 2016. Biomagnetic monitoring through *Lantana* leaves in an Indo-Burma hot spot region. *Environ Skeptics Critics* 5(1):1–11.

Rai PK, Mishra A, Tripathi BD. 2010. Heavy metals and microbial pollution of river Ganga: A case study on water quality at Varanasi. *Aquat Ecosyst Health Manag* 13(4):352–361.

Rai PK, Singh MM. 2015. *Lantana camara* invasion in urban forests of an Indo-Burma hotspot region and its ecosustainable management implication through biomonitoring of particulate matter. *J Asia Pac Biodivers* 8:375–381.

Rai PK, Singh MM. 2016. *Eichhornia crassipes* as a potential phytoremediation agent and an important bioresource for Asia Pacific region. *Environ Skeptics Critics* 5(1):12–19.

Seidel K. 1964. Abbau von bacterium coli durch höhere wasserpflanzen. *Naturwissenschaften* 51:395.

Seidel K. 1976. Macrophytes and water purification. In *Biological Control of Water Pollution*, ed. J Tourbier, RWJ Pierson, 109–123. Philadelphia: University of Pennsylvania Press.

Shannon CE, Weaver W. 1949. *The Mathematical Theory of Communication*. Urbana: University of Illinois Press.

Simpson EH. 1949. Measurement of diversity. *Nature* 163:688.

Singh MM, Rai PK. 2016. Microcosm investigation of Fe (iron) removal using macrophytes of Ramsar lake: A phytoremediation approach. *Int J Phytoremediation* 18(12):1231–1236.

Sorenson T. 1948. A method of establishing groups of equal amplitude in plant sociology based on similarity of species content. K. *Dans ka Videnskab Selsk Biol Skr* 5(4):1–34.

Tsang E. 2015. Effectiveness of wastewater treatment for selected contaminants using constructed wetlands in Mediterranean climates. Master's projects, University of San Francisco.

Veenstra S. 1998. The Netherlands. In *Constructed Wetlands for Wastewater Treatment in Europe*, ed. J Vymazal, H Brix, PF Cooper, MB Green, R Haberl, 289–314. Leiden, the Netherlands: Backhuys Publishers.

Vymazal J. 2006. Removal of nutrients in various types of constructed wetlands. *Sci Total Environ* 380(1–3):48–65.

WaterAid. 2008. *Decentralized wastewater management using constructed wetlands in Nepal*. London: WaterAid.

Wieder R. 1989. A survey of constructed wetlands for acid coal mine drainage treatment in the eastern United States. *Wetlands* 9(2):299–315.

Zhang BY, Zheng JS, Sharp RG. 2010. Phytoremediation in engineered wetlands: Mechanisms and applications. *Procedia Environ Sci* 2:1315–1325.

6 Global Ramsar Wetland Sites: A Case Study on Biodiversity Hotspots

GLOBAL RAMSAR SITES AND NATURAL WETLANDS OF THE TROPICAL AND TEMPERATE WORLD

Ecosystem services or environmental management role of global wetlands are immense (Nahlik and Mitsch, 2006; Maltby and Acreman, 2011; Mbuligwe et al., 2011; Verma and Negandhi, 2011). Further, global wetlands alleviate the socioeconomic status of an existing region and assist in attaining sustainable development (Verma and Negandhi, 2011). Statistically, wetlands cover 6% of the Earth's total surface area (The Ramsar Convention on Wetlands 2013; Feng et al., 2013; Pour et al., 2015). They have a total area of 885 million ha, 190 million ha of which has been recorded in the Ramsar Convention (Pour et al., 2015).

Iran is the country where the Ramsar wetland concept was innovated; however, little research has been performed in this country's wetland itself. A study assessed the status of wetlands in the basin of the Caspian Sea (including Anzali, Boojagh, and Miyankaleh wetlands at the southern border) as an ecosystem with unique ecological features through the Caspian Rapid Assessment Method (Pour et al., 2015). Shadegan wetland (consisting of wetland plants like *Phragmites australis*, *Typha australis*, and *Scripus maritimus*) in southwest Iran witnessed a bioaccumulation of emerging contaminants/trace metals in wetland plants, sediments, and greater concentrations in birds or fishes (Alhashemi et al., 2011, 2012). Likewise, the Anzali wetland (a Ramsar site of Iran), with the help of a submerged aquatic plant *Ceratophyllum demersum*, removed several metallic contaminants like cadmium (Cd), copper (Cu), zinc (Zn), lead (Pb), and chromium (Cr) (Pourkhabbaz et al., 2011).

In Egypt, *Arundo donax*, a potent wetland plant, showed its potential for emerging contaminant eco-removal, particularly in the context of heavy metals (Cd) and nutrients (Galal and Shehata, 2016). Further, in an Asian country's (Pakistan) context, this wetland plant (*A. donax*), through volatilization mechanism (as discussed in previous chapters), remediated soil, specifically, arsenic, a metalloid of particular health concern (Mirza et al., 2011).

It has been discussed that Fe plaques are formed under metal stress in global wetland plants, which assist in the removal of emerging contaminants like arsenic. Nevertheless, arsenic entrapment in Fe plaque may pose an eco-toxicological risk to wetland herbivores, and thus, special caution is needed in this context (Taggart et al., 2009).

In Mexico, Chimaliapan wetland (a Ramsar site) was also investigated for the impacts of emerging contaminants on life forms, specifically zooplanktons

(García-García et al., 2012). In Tunisia, the wetland in El Kelbia sebkhet has been investigated for its water quality (Duplay et al., 2013). In Hungary, the well-planned development of wetlands and geographical changes in the past have been investigated due to their extreme environmental relevance (Uj et al., 2016).

In Hong Kong, Mai Po, a Ramsar wetland, is deeply polluted with emerging contaminants like polycyclic aromatic hydrocarbons, and several bioassay studies have been done, which may be replicated in other wetlands (Kwok et al., 2010). In a natural wetland in Phnom Penh, Cambodia, x-ray fluorescence greatly assisted in emerging contaminants (heavy metals) associated with sediment (Sereyrath et al., 2016). Furthermore, in an important urban coastal wetland in Ghana, the emergent aquatic macrophytes or wetland plants *Typha domingensis*, *Ludwigia* sp., and *Paspalum vaginatum* respectively had the highest accumulation capacity for Cd, As, and Hg, but the floating aquatic plant *Pistia stratiotes* appeared to be a better accumulator of Cd and As (Gbogbo and Otoo, 2015). In Northern Greece, Agras wetland was found to have healthy relationship between hydrochemistry, hydrodynamics, habitats, and human interventions (Grigoriadis et al., 2009).

RAMSAR SITES IN INDIA AND PHYTOREMEDIATION WORK

India has a varying topography and climatic regimes and supports diverse and unique wetland habitats (Prasad et al., 2002). To date, 201,503 wetlands in India have been identified and mapped on a 1:50,000 scale (SAC, 2011). Overall, India has about 757.06 thousand wetlands of 15.3 m ha (nearly 4.7% of the total geographical area), which include open water, aquatic vegetation (submerged, floating, and emergent), and surrounding hydric soils (Bassi et al., 2014). These wetlands provide numerous products and services to humanity with their unique ecological features (Prasad et al., 2002).

Ramsar sites, as per the Ramsar convention, defined as most of the natural bodies such as rivers, lakes, coastal lagoons, mangroves, peat land, coral reefs, etc., constitute the wetland ecosystem. In India, there are 26 Ramsar sites or wetlands of international importance (Ramsar Convention Secretariat, 2013). Phytoremediation work in India has been done by several researchers and scientists. In Ramsar sites, the Hokersar wetlands of Kashmir Himalaya were studied to assess the heavy-metal, such as Al, Mn, Ba, Zn, Cu, Pb, Mo, Co, Cr, Cd, and Ni, sequestration capability of the macrophyte species *P. australis* (Ahmad et al., 2014). The East Calcutta Wetlands of Kolkata were also studied for waste metal, such as Ca, Cr, Mn, Fe, Cu, Zn, and Pb, remediation using floriculture of the plant species such as *Helianthus annuus*, *Tagetes patula*, and *Celocia cristata* (Chaterjee, 2011). In Northeast India, the three Ramsar sites Deepor Beel of Assam, Loktak Lake of Manipur, and Rudrasagar lake of Tripura have not been studied on phytoremediation. The current study will be focusing on the phytoremediation work using the local macrophytes in the Loktak Lake of Manipur, which is the largest freshwater lake in eastern India.

Emerging contaminants from diverse industrial effluents are reported to pollute the East Calcutta Wetlands, a Ramsar site in West Bengal, India (12,500 ha area), of much ecological and economic relevance (Chaterjee et al., 2007, 2011). To this end, a wetland plant water hyacinth (*Eichhornia crassipes*) was found to be efficient in eco-removal/phytoremediation of emerging contaminants, which was analyzed in

E. crassipes, wetland system, and fishes with the help of an advanced instrument, that is, particle-induced x-ray emission (Chaterjee et al., 2007). Further, Bermuda grass (*Cynodon dactylon*) was found to be efficient in eco-removal/phytoremediation of emerging contaminants like metals in East Calcutta Wetlands, a Ramsar site in West Bengal, India (Chaterjee et al., 2011). In the same wetland, i.e., East Calcutta Wetlands, a phytoremediation study based on *Colocasia esculenta* and *Scirpus articulates* was performed successfully for heavy metals (Khatun, 2016). Another mangrove wetland, i.e., Indian Sundarban Wetland, has been assessed recently for trace metal accumulation in several wetland plants (Rodríguez-Iruretagoiena et al., 2016).

To this end, a couple of submerged plants, *Potamogeton–Ceratophyllum* combination, proved to be useful in maintaining a healthy ecosystem of a Himalayan Ramsar wetland (Hokersar wetland) through removal of emerging contaminants, especially metals (Ahmad et al., 2016). Further, *P. australis* in Hokersar wetland was also useful in the removal of emerging contaminants, especially metals, thus protecting the other life forms of this important wetland (Ahmad et al., 2014). Kolleru Lake (a Ramsar wetland in Andhra Pradesh, India) was investigated for its water quality (water chemistry and sediment ^{13}C and ^{18}O compositions), and findings may be useful for environmental rehabilitation of the lake and its surroundings (Sharma and Sujatha, 2016). In India, the Upper Ganga Ramsar site of Uttar Pradesh, consisting of free-floating *E. crassipes*, *P. stratiotes*, and *Lemna minor*, effectively removed emerging contaminants/heavy metals (Garg and Joshi, 2015).

LOKTAK LAKE: AN IMPORTANT RAMSAR WETLAND

Wetlands are defined as "lands transitional between terrestrial and aquatic ecosystems where the water table is usually at or near the surface or the land is covered by shallow water" (Mitsch and Gosselink, 1986). Nowadays, wetlands are fast declining and rapidly deteriorating ecosystems in the world. People around the world will have to make concerted efforts for the abatement toward the degradation of lakes. Fresh water lakes are of much importance to mankind, but they occupy a relatively small portion of the earth's surface as compared to the marine and terrestrial habitats (Santra, 2001). The Loktak Lake, the largest fresh water lake in Northeast India, is rich in biodiversity, and it is considered to be the "lifeline for the people of Manipur" due to its importance in their socioeconomic and cultural life (Tombi and Shyamananda, 1994) and has been recognized as a Wetland of International Importance (Ramsar site no. 463, declared on June 16, 1993). The "phoomdi" (a Manipuri word meaning floating mats of soil and vegetation) is a heterogeneous mass of soil, vegetation, and organic matter in different stages of decay, which has a unique ecosystem, and it is only found in this particular lake.

DESCRIPTION OF THE STUDY SITE OF RAMSAR SITE OF GLOBAL BIODIVERSITY HOTSPOT

Manipur is one of the eight states of Northeast India. The state is bound by Nagaland in the north, by Mizoram in the south, Assam in the west, and the borders of Burma in the east as well as in the south. The state capital of Manipur is Imphal. The state lies at latitude of 23°83′N–25°68′N and longitude of 93°03′E–94°78′E. The total area covered by the state is 22,347 km². The capital lies in an oval-shaped valley of approximately 700 square miles (2000 km²), surrounded by blue mountains, and is at an elevation of 790 m above sea level. The slope of the valley is from north to south. The presence of mountain ranges not only prevents the cold winds from the north from reaching the valley but also acts as a barrier to the cyclonic storms originating from the Bay of Bengal. The southwest monsoon chiefly determines the weather and rainfall throughout the state. The state has a tropical to temperate climate depending upon elevation. Rainfall varies from 1000 to 3500 mms, and average rainfall is 1500 mms. Temperature ranges from subzero to 36°C. Winter season is from November to February. Then, the premonsoon falls during the months of March and April. The monsoon season is from the month of May to September, and the postmonsoon season is in October and November.

In the heart of Manipur, there lies the Loktak Lake (Map 6.1), which is rich in biodiversity and considered to be the lifeline of Manipur valley and has been recognized as a Wetland of International Importance (Ramsar site no. 463, declared on March 23, 1990), which was added in the Montreux Record on June 16, 1993. The Loktak Lake lies between the latitude of 24°25′–24°42′N and longitude of 93°46′–93°55′E, located at the Bishnupur district of Manipur, and is the largest natural lake in eastern India; its size is approximately 26,600 ha. The lake also supports varied types of habitat, due to which the lake is blessed with a rich diversity of flora and fauna. The Keibul Lamjao National Park, which is approximately 40.5 km², is the largest among the phoomdis in the Loktak Lake, which is home to the Sangai (*Rucervus eldii eldii*), the Manipuri brow-antlered deer, which is on the brink of extinction. The lake sustains the lives of macrophytes, the wildlife and people who live on the phoomdis (Ningombam and Bordoloi, 2008).

SELECTION OF SAMPLING SITES

Preliminary surveys of the lake were done before selection of the sampling sites for detailed investigation in relation to the physicochemical and phytosociological studies of the water. A total of four sampling sites were selected (in triplicates) for analysis of various physicochemical characteristics of the water. The study sites (Map 6.2) were selected on the basis of the source pollution and disturbance causing the lake to be polluted. The sites are as follows.

1. Site I (Loktak Nambul vicinity):
   Site I is the contact point of Loktak Lake and the river, Nambul, which passes through the Imphal municipal area and then drains in the lake. This river carries waste of the Imphal municipal area.
2. Site II (Loktak Nambol vicinity):
   Site II is the contact point of the Loktak Lake and the river, Nambol, which passes through the Bishenpur municipal area and then drains in the lake. This river carries waste of the Bishenpur municipal area.

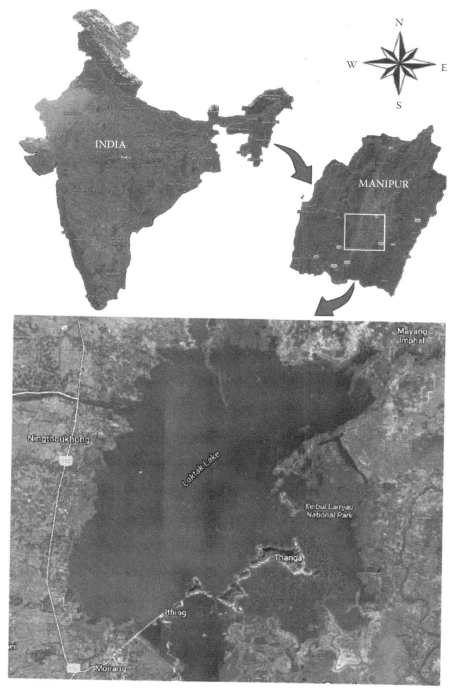

MAP 6.1   Loktak Lake at Bishenpur District, Manipur.

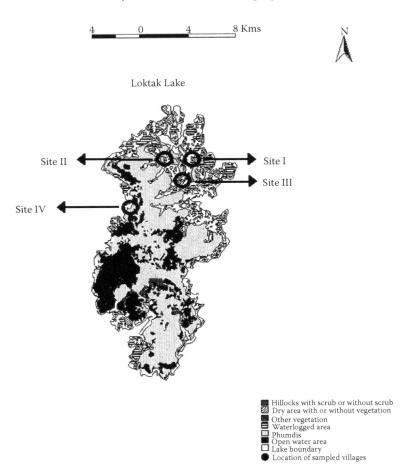

MAP 6.2   Structure of the Loktak Lake and study sites.

3. Site III (Loktak Yangoi vicinity):
    Site III is the contact point of the Loktak Lake and the river, Yangoi,
    which is also a main source of pollution drained in the lake.
4. Site IV (Loktak proper):
    Site IV is the Joining point of the Loktak Lake and the Canal of National
    Hydro Power Corporation Limited of Loktak Lake (photo Plates 6.1–6.4).

Water sampling was performed in all three seasons, i.e., rainy season, winter
season, and summer season, from August 2013 to July 2015.

DISTRIBUTION

India has 2167 recorded natural wetlands, covering an area of 1.5 million hectares.
Further, there are 65,254 artificial wetlands, spreading over an area of 0.25 million
hectares (Kumar, 1999). Indian wetlands cover the whole range of the ecosystem

PHOTO PLATE 6.1   Study site I. (Courtesy of M. Muni Singh, author's research student.)

PHOTO PLATE 6.2   Study site II. (Courtesy of M. Muni Singh, author's research student.)

PHOTO PLATE 6.3   Study site III. (Courtesy of M. Muni Singh, author's research student.)

PHOTO PLATE 6.4   Study site IV. (Courtesy of M. Muni Singh, author's research student.)

types found, including natural wetlands of the high-altitude Himalayan lakes, followed by wetlands of flood plains of the major river systems, saline wetlands, and wetlands of the arid and semi-arid regions, coastal wetlands such as lagoons and estuaries; mangrove swamps; coral reefs; marine wetlands; and so on (Prasad et al., 2002). The biodiversity-rich Northeast India accounts for three Ramsar sites in the country, i.e., Deepor Beel in Assam, Rudra Sagar in Tripura, and Loktak Lake in Manipur, being the major wetland-areas in the region.

VALUES

Loktak Lake performs numerous valuable functions. It recycles nutrients; purifies water; recharges ground water; provides drinking water, fish, fodder, fuel, and wildlife habitat; controls the rate of runoff in urban areas; and acts as a recreation center in the state. The interaction of humans with wetlands during the last few decades has been of concern largely due to the rapid population growth, accompanied by intensified industrial, commercial, and residential development, further leading to pollution of wetlands by domestic, industrial sewage, and agricultural run-offs as fertilizers, insecticides, and feedlot wastes (Prasad et al., 2002). The fact that wetland values are overlooked has resulted in threat to the source of these benefits. Wetlands are often described as "kidneys of the landscape" (Mitsch and Gosselink, 1986). Hydrologic conditions can directly modify or change chemical and physical properties such as nutrient availability, degree of substrate anoxia, soil salinity, sediment properties, and pH. These modifications of the physiochemical environment, in turn, have a direct impact on the biotic response in the wetland (Gosselink and Turner, 1978).

DIVERSITY AND RESOURCES

The lake supports varied types of habitat, due to which the lake is blessed with a rich diversity of flora and fauna. The lake sustains the lives of the wildlife (flora and fauna) and people who live on the phoomdis.

Animal Resources

The lake is the natural habitat of the most endangered ungulate species, i.e., the brow antlered deer (*Rucervus eldii eldii*), locally known as Sangai (Kosygin and Dhamendra, 2009). The Keibul Lamjao National Park, which is approximately 40.5 km², is the largest among the phoomdis in the lake and is home to the Sangai. A total of 425 species of animals (249 vertebrates and 176 invertebrates) have been spotted in the lake. Rare animals like Indian python, samber and barking deer, *Muntiacus muntjak*, rhesus monkey, hoolock gibbon, hog deer, otter, wild boar, fox, jungle cat, golden cat, sambar, etc., are found in the lake. A total of 116 species of birds, including 21 species of waterfowl, have been spotted in the lake. Varied species of water fowl and migratory birds could be spotted during November to March. It is also an important bird area as it is being a potential breeding site for some waterfowl and is a staging site for migratory birds, especially from Siberia. The prominent bird species found are east Himalayan pied kingfisher, hornbills, black kite, lesser skylark, northern hill myna, lesser eastern jungle crow, yellow headed wagtail, spotbill duck, and Indian white breasted water hen, among others. Among the vertebrates fishes, *Channa striatus* and *Channa punctatus* are found in the park. Fish yield from the lake is reported to be 1500 tons every year. Amphibians and reptiles in the park include keel black tortoise, viper, krait, Asian rat snake, and python (Singh and Rai, 2014; Rai and Singh, 2014). Due to anthropogenic activities like agricultural expansion and draining of polluted water to the lake from sewage as well as industries adversely affected the faunal population in vicinity of this lake. It specially affects the Sangai (indigenous deer found only in Loktak Lake) keeping below the near extinct status.

Plant Resources

Ecological studies of Loktak Lake reported that, altogether, 86 macrophytic plant species are found distributed in the lake in different seasons of the year. *E. crassipes*, *Euryale ferox*, *Nelumbo nucifera*, *Nymphea pubescence*, *Nymphoides indicum*, and *Trapa natans* were the most common species. Thirteen macrophyte species, i.e., *C. demersum*, *E. crassipes*, *E. ferox*, *Hydrilla verticillata*, *Nymphoides cristatum*, *P. stratiotes*, *Potamogeton crispus*, *Salvinia cucullata*, *Salvinia natans*, *T. natans*, *Urticularia exoleta*, *Urticularia flexuosa*, and *Vallisnaria spiralis*, are found distributed in all months of the year (Devi and Sharma, 2002). Important vegetation of the phoomdis includes *E. crassipes*, *Phragmites karka*, *Oryza sativa*, *Zizania latifolia*, *Cynodon* spp., *Limnophila* spp., *Sagittaria* spp., *Saccharum latifolium*, *Erianthus pucerus*, *Erianthus ravennae*, *Lersia hexandra*, *Carex* spp., etc., which are the most dominant species, and floating plants include *N. nucifera*, *T. natans*, *E. ferox*, *Nymphaea alba*, *Nymphaea nouchali*, *Nymphaea stellata*, and *Nymphoides indica* (Sanjit, 2005).

Species of economic importance like *T. natans*, *E. ferox*, *N. nucifera*, and *Nympheae* spp. have been found to be greatly decreased in growth and production over the year.

Depleting Water Quality

The last few decades have been a very critical time in the degradation of Loktak Lake, due to the expansion of agricultural activities. This is due to the activities of the people inhabiting around the lake and also to some rivers that are drained into Loktak Lake. The rivers that drain directly to the lake, i.e., Potsangbam, Awang khujairok, Thongjarok, Merakhong, Nambol, and Nambul, mostly contain a heavy load of agricultural chemicals as well as domestic waste from different sources in the Imphal city into the lake water and may contribute significantly to the water quality deterioration of the lake; if this is not taken into consideration, it may also result in eutrophication of the lake (Loktak Development Authority, 2011). As a result, there is an enormous increase in the production of unwanted plants, leading to an increase in phoomdis. The lake is also gradually silting, and the pollution of its water is increasing day by day due to those activities that lead to the shrinkage of the lake (Roy, 1992). Direct discharge of urban waste from Imphal city into Loktak Lake through Nambul River has become one of the main causes for polluting the lake. Some expected heavy-metal contamination might occur in the lake due to the excessive discharge of waste from Imphal city in the river, especially Nambul River, which, after draining into the lake directly as heavy metals, is known to have an adverse or detrimental effect on the environment and human health by their carcinogenic effect. Some elements were detected in the lake water, such as Mg, Al, P, S, Cl, K, Ca, Mn, Fe, Cu, Zn, Se, Br, Rb, and Sr, some of which were detected with high concentration, leading to the fact that there is some complex relationship between environmental concentrations and bioaccumulation (Singh et al., 2013). Internally, Loktak Lake is weakening and becoming externally disturbed, with a dismal future in all manifestations. The water quality of Loktak Lake, in general, falls within class C to E as per the Central Pollution Control Board (CPCB) designated best use criteria, and the lake water is not fit for direct drinking without treatment but can be used for irrigation purposes (Rai and Raleng, 2011).

PHYTOREMEDIATION IN THE PRESENT CONTEXT OF BIODIVERSITY HOTSPOT

The use of plants for the removal of pollutants from the environment is termed phytoremedition (Gardea Torresdey, 2005). Phytoremediation refers to the natural ability of certain plants called hyperaccumulators to bioaccumulate, degrade, or render harmless contaminants in soils, water, or air (Rai, 2007a,b, 2008a–c, 2009a,b, 2012, 2013, 2015a–d, 2016; Rai and Singh, 2015, 2016; Rai and Chutia, 2016; Singh and Rai, 2016). Phytoremediation is an emerging technology that uses various plants to degrade, extract, contain, or immobilize contaminants from soil and water. This technology has been receiving attention lately as an innovative, cost-effective alternative to the more established treatment methods used at hazardous waste sites. The huge cost burden has opened a path to the marketplace for an innovative technology. The removal of heavy metals using living organisms has recently been attracting a lot of public attention and research and development spending. The expansion of this research work will promote the eco-friendly and cost-effective technology as phytoremediation. Phytoremediation promotes the use of plants for environmental cleanup.

Role of Phytoremediation in a Ramsar Site of Biodiversity Hotspot

The use of phytoremediation as an abatement of water pollution has been studied by many researchers. Researchers used many macrophytic plants to study the accumulation of metals to levels greatly in excess of these in their environment. Macrophytes such as *E. crassipes, Azolla pinnata, H. verticillata, L. minor, Ipomoea aquatica, Marsilea quadrifolia, Vallisneria spiralis*, etc., were used for the abatement of the pollution. *C. demersum, Echinochloa pyramidalis, E. crassipes, Myriophyllum spicatum, P. australis*, and *T. domingensis* can survive in extreme conditions and can tolerate very high concentrations of heavy metals, which make them an excellent choice for phytoremediation and biomonitoring programs (Fawzy et al., 2012; Singh and Rai, 2016). Aquatic plants absorb elements through roots and/or shoots (Pip and Stepaniuk, 1992). In aquatic systems, where pollutant inputs are discontinuous and pollutants are quickly diluted, analyses of plant tissues provide time-integrated information about the quality of the system (Baldantoni et al., 2004, 2005). Exhaustive monitoring and assessment of heavy-metal pollution and phytoremediation experiments through diverse macrophytes were performed in Singrauli Industrial Region, India (Rai, 2008a–c; Rai and Tripathi, 2009), and subsequently, human health implications were assessed indirectly (Rai, 2008b). The use of aquatic macrophytes, such as *A. pinnata*, with hyperaccumulating ability is known to be an environmentally friendly option to restore polluted aquatic resources (Sood et al., 2012). Once the pollutants are released to the water bodies, the plants are the only hope to clean up the pollutants by absorbing and metabolizing them from the water bodies. Therefore, the role of plants in water pollution abatement is very important in the present era of rapid industrialization and urbanization.

A phytoremediation technique using the plant species found in the lake will help in the prevention of further pollution of the water, including heavy-metal pollution, and will increase the proper knowledge of utilizing the particular plant species, which is very important for the wetland and for biodiversity conservation of this lake.

There are little research that has studied physicochemical characteristics, heavy metal, phytosociology, as well as phytoremediation studies in totality. Henceforth, the present study aimed

- To assess the heavy-metal pollutants of Loktak Lake.
- To perform phytosociological studies of macrophytes.
- To perform heavy-metal phytoremediation investigations with selected macrophytes of the Loktak Lake.

WATER QUALITY ANALYSIS IN A RAMSAR SITE OF BIODIVERSITY HOTSPOT

The analysis for water quality parameters was conducted for two years, i.e., from August 2013 to July 2015, on a seasonal basis considering the mean values for each water quality parameter. The results of the various water quality parameters of the four sites, with standard deviation, with seasonal variation are given in Figures 6.1 through 6.13.

TEMPERATURE

The highest value of temperature was measured as 28.6°C at site II during the rainy season (2013–2014) and the lowest was measured as 16.8°C at site II during the winter season (2013–2014), as shown in Figure 6.1. Rainy seasons normally possess a higher temperature; on the other hand, the winter seasons possess a lower temperature. All the water quality parameters relating to the physicochemical characteristics from different seasons were grouped together site wise and were correlated to find the correlation coefficient between them (Tables A.1 through A.4).

pH

The highest value of pH was measured as 7.4 at site I and site IV during the summer season (2014–2015), and the lowest was 6.3 at site IV during the rainy season (2013–2014), as shown in Figure 6.2. The values for site I and site IV, i.e., 6.4 and 6.3, are lower than the permissible limit value set by the World Health Organization (WHO) and Indian Council of Medical Research (ICMR). The summer season has the highest pH value, whereas the rainy season has the lowest pH. All the water quality parameters relating to the physicochemical characteristics from different seasons were grouped together site wise and were correlated to find the correlation coefficient between them (Tables A.1 through A.4).

TRANSPARENCY

The highest value of transparency was measured at 1.7 m at site IV during the winter season (2013–2014) and the rainy season (2013–2014), and the lowest value of 0.74 m was measured at site III during the summer season (2013–2014), as shown in Figure 6.3. The winter season has the highest transparency, and the summer season has the lowest transparency of the water. All the water quality

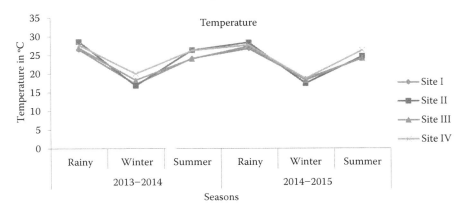

FIGURE 6.1   Seasonal variations of temperature of water from different study sites.

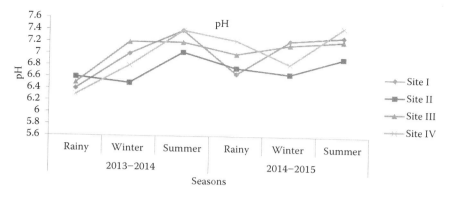

FIGURE 6.2   Seasonal variations of pH of water from different study sites.

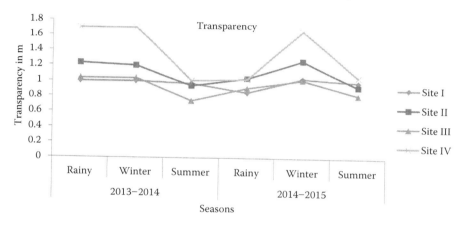

FIGURE 6.3   Seasonal variations of transparency of water from different study sites.

parameters relating to the physicochemical characteristics from different seasons were grouped together site wise and were correlated to find the correlation coefficient between them (Tables A.1 through A.4).

TOTAL SOLIDS

The highest value of total solids (TS) was measured as 46.3 mg L^{-1} at site III during the rainy season (2014–2015), and the lowest value of 23.3 mg L^{-1} was measured at site III during the winter season (2013–2014) and site I and site III during the summer season of both the years 2013–2014 and 2014–2015, as shown in Figure 6.4. Normally, due to agitation of water with the rains, the rainy season has the highest TS and the winter season has the lowest TS. All the water quality parameters relating to the physicochemical characteristics from different seasons were grouped together site wise and were correlated to find the correlation coefficient between them (Tables A.1 through A.4).

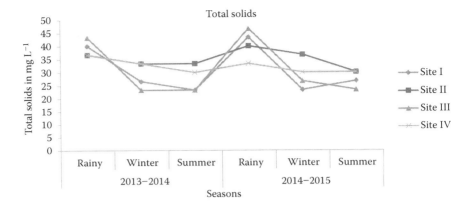

FIGURE 6.4   Seasonal variations of TS of water from different study sites.

Dissolved Oxygen

The highest dissolved oxygen (DO) was measured as 9.6 mg L^{-1} at site IV during the winter season (2013–2014), and the lowest value of 6.9 mg L^{-1} was measured at site III during the summer season (2013–2014), as shown in Figure 6.5. The winter season has the highest DO, and the summer season has the lowest DO. All the water quality parameters relating to the physicochemical characteristics from different seasons were grouped together site wise and were correlated to find the correlation coefficient between them (Tables A.1 through A.4).

Biological Oxygen Demand

The highest value for biological oxygen demand (BOD) was measured as 2.1 mg L^{-1} at site I during the winter season (2014–2015), and the lowest value of 0.5 mg L^{-1} was measured at site III during the winter season (2013–2014), as shown in Figure 6.6.

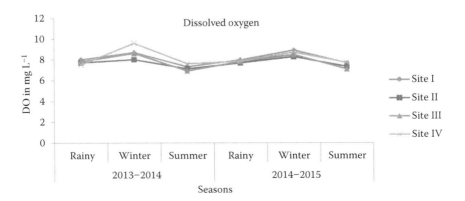

FIGURE 6.5   Seasonal variations of DO of water from different study sites.

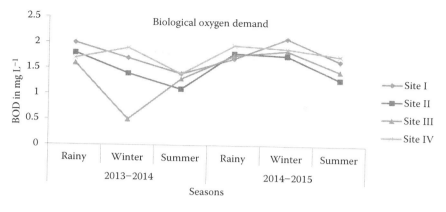

FIGURE 6.6 Seasonal variations of BOD of water from different study sites.

Winter season has the highest BOD and the winter season and rainy season has the lowest BOD. All the water quality parameters relating to the physicochemical characteristics from different seasons were grouped together site wise and were correlated to find the correlation coefficient between them (Tables A.1 through A.4).

ACIDITY

The highest value of acidity was measured as 32.6 mg L^{-1} at site III during the rainy season (2013–2014), and the lowest value of 9.3 mg L^{-1} was measured at site IV during the winter season (2013–2014), as shown in Figure 6.7. Rainy seasons have the highest acidity, and winter seasons have the lowest. All the water quality parameters relating to the physicochemical characteristics from different seasons were grouped together site wise and were correlated to find the correlation coefficient between them (Tables A.1 through A.4).

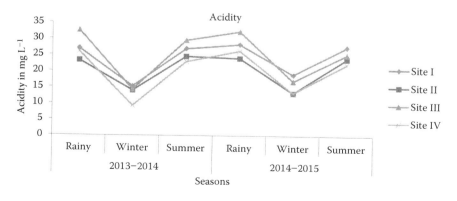

FIGURE 6.7 Seasonal variations of acidity of water from different study sites.

ALKALINITY

The highest value of alkalinity was measured as 73.3 mg L^{-1} at site IV during the winter season (2013–2014), and the lowest value of 21.3 mg L^{-1} was measured at site III during the rainy season (2013–2014), as shown in Figure 6.8. Winter seasons have the highest alkalinity, and rainy seasons have the lowest. All the water quality parameters relating to the physicochemical characteristics from different seasons were grouped together site wise and were correlated to find the correlation coefficient between them (Tables A.1 through A.4).

CHLORIDE

The highest value of chloride was measured as 45.6 mg L^{-1} at site III during the winter season (2014–2015), and the lowest value of 20.3 mg L^{-1} was measured at site II during the winter season (2014–2015), as shown in Figure 6.9. Winter has the highest chloride as well as lowest. All the water quality parameters relating to

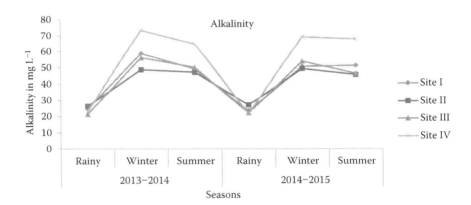

FIGURE 6.8 Seasonal variations of alkalinity of water from different study sites.

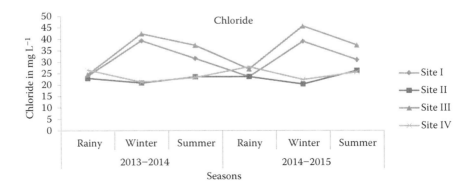

FIGURE 6.9 Seasonal variations of chloride of water from different study sites.

the physicochemical characteristics from different seasons were grouped together site wise and were correlated to find the correlation coefficient between them (Tables A.1 through A.4).

TOTAL HARDNESS

The highest value of total hardness was measured as 68 mg L^{-1} at site III during the winter season (2013–2014), and the lowest value of 18.7 mg L^{-1} was measured at site III during the rainy season (2013–2014), as shown in Figure 6.10. The winter season has the highest total hardness value, whereas the rainy season has the lowest. All the water quality parameters relating to the physicochemical characteristics from different seasons were grouped together site wise and were correlated to find the correlation coefficient between them (Tables A.1 through A.4).

TURBIDITY

The highest value of turbidity was measured as 15.8 NTU at site IV during the summer season (2013–2014), and the lowest value of 0.4 NTU was measured at site III (2013–2014) and site IV (2013–2014 and 2014–2015) during the winter season, as shown in Figure 6.11. The summer season has the highest turbidity, and the winter season has the lowest. Some values during the summer and rainy seasons are higher than the permissible limits set by WHO. All the water quality parameters relating to the physicochemical characteristics from different seasons were grouped together site wise and were correlated to find the correlation coefficient between them (Tables A.1 through A.4).

NITRATE

The highest value of nitrate was found as 0.62 mg L^{-1} at site II during the summer season (2013–2014), and the lowest value of 0.21 mg L^{-1} was measured at site IV

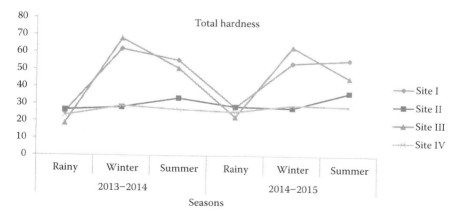

FIGURE 6.10 Seasonal variations of total hardness of water from different study sites.

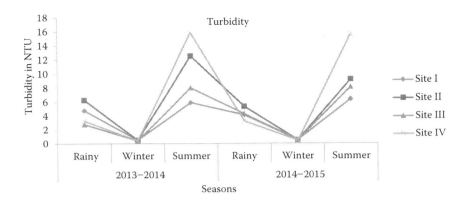

FIGURE 6.11   Seasonal variations of turbidity of water from different study sites.

during the rainy season (2013–2014), as shown in Figure 6.12. The summer season has the highest nitrate, and the winter season has the lowest. All the water quality parameters relating to the physicochemical characteristics from different seasons were grouped together site wise and were correlated to find the correlation coefficient between them (Tables A.1 through A.4).

PHOSPHATE

The highest value of phosphate was measured as 0.3 mg L^{-1} at site II during the summer season (2014–2015), and the lowest value of 0.06 mg L^{-1} was measured at site IV during the winter season (2013–2014), as shown in Figure 6.13. The summer season has the highest phosphate, and the winter season has the lowest. Phosphate values were higher than the permissible limit of United States Public Health Service (USPH). All the water quality parameters relating to the physicochemical characteristics from different seasons were grouped together site wise and were correlated to find the correlation coefficient between them (Tables A.1 through A.4).

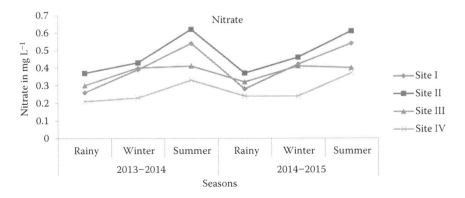

FIGURE 6.12   Seasonal variations of nitrate of water from different study sites.

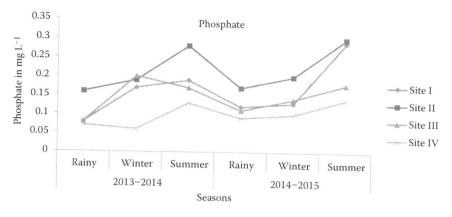

FIGURE 6.13 Seasonal variations of phosphate of water from different study sites.

HEAVY-METALS ANALYSIS

The analysis for heavy metals for the water samples collected from the different sites was conducted for two years, i.e., from August 2013 to July 2015, on a seasonal basis, as well for the macrophytes collected from the different sites but not seasonally. Seven heavy metals, i.e., iron (Fe), mercury (Hg), cadmium (Cd), arsenic (As), lead (Pb), chromium (Cr), and zinc (Zn), were analyzed for the water samples as well as for the macrophytes. The macrophytes collected from the different sites are *E. crassipes*, *L. minor*, *P. stratiotes*, and *S. cucullata*. The results of the heavy-metal concentrations of the four sites for water and plants with seasonal variations are given in Figure 6.14 and Tables 6.1 through 6.6.

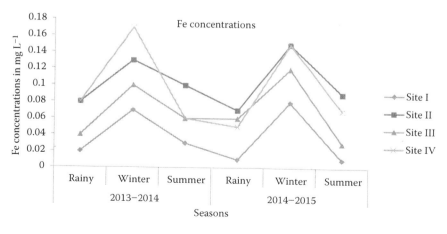

FIGURE 6.14 Seasonal variations of Fe concentrations (in mg L^{-1}) of water from different study sites.

WATER

Water samples were collected in all the seasons for all the four sites. The results are as follows.

Iron (Fe)

The highest value of Fe concentration was measured as 0.17 mg L^{-1} at site IV during the winter season (2013–2014), and the lowest value of 0.01 mg L^{-1} was measured at site I during the rainy season (2014–2015) and summer season (2014–2015), as shown in Figure 6.14. The winter season has the highest Fe concentration, and the rainy season has the lowest.

Mercury (Hg)

Hg was present in a minute negligible amount of <0.001 mg L^{-1} for all the samples collected from the different sites, as shown in Table 6.1. In this study, all the values of Hg concentrations are found far below the limit and are not harmful to use for any purposes.

Cadmium (Cd)

Cd was present in a minute negligible amount of <0.001 mg L^{-1} for all the samples collected from the different sites, as shown in Table 6.2. In this study, all the values of Cd concentrations are found far below the limit and are not harmful to use for any purposes.

TABLE 6.1
Hg Concentrations (in mg L^{-1}) of Water from Different Study Sites

Study Sites	2013–2014			2014–2015		
	Rainy	Winter	Summer	Rainy	Winter	Summer
Site I	<0.001	<0.001	<0.001	<0.001	<0.001	<0.001
Site II	<0.001	<0.001	<0.001	<0.001	<0.001	<0.001
Site III	<0.001	<0.001	<0.001	<0.001	<0.001	<0.001
Site IV	<0.001	<0.001	<0.001	<0.001	<0.001	<0.001

TABLE 6.2
Cd Concentrations (in mg L^{-1}) of Water from Different Study Sites

Study Sites	2013–2014			2014–2015		
	Rainy	Winter	Summer	Rainy	Winter	Summer
Site I	<0.001	<0.001	<0.001	<0.001	<0.001	<0.001
Site II	<0.001	<0.001	<0.001	<0.001	<0.001	<0.001
Site III	<0.001	<0.001	<0.001	<0.001	<0.001	<0.001
Site IV	<0.001	<0.001	<0.001	<0.001	<0.001	<0.001

Arsenic (As)

As was present in a minute negligible amount of <0.001 mg L^{-1} for all the samples collected from the different sites, as shown in Table 6.3. In this study, all the values of As concentrations are found far below the limit and are not harmful to use for any purposes.

Lead (Pb)

Pb was present in a minute negligible amount of <0.001 mg L^{-1} for all the samples collected from the different sites, as shown in Table 6.4. In this study, all the values of Pb concentrations are found far below the limit and are not harmful to use for any purposes. Pb is a cumulative poison, and its effects on human health include gastrointestinal disorder, liver and kidney damage, abnormalities, and infertility. Pb poisoning is due to permanent cumulative effects and not due to occasional exposure to small doses. However, in extreme case of Pb poisoning, death may result (Trivedy and Goel, 1986).

Chromium (Cr)

Cr was present in a minute negligible amount of <0.001 mg L^{-1} for all the samples collected from the different sites, as shown in Table 6.5. In this study, all the values of Cr concentrations are found far below the limit and are not harmful to use for any purposes.

TABLE 6.3
As Concentrations (in mg L^{-1}) of Water from Different Study Sites

Study Sites	2013–2014			2014–2015		
	Rainy	Winter	Summer	Rainy	Winter	Summer
Site I	<0.001	<0.001	<0.001	<0.001	<0.001	<0.001
Site II	<0.001	<0.001	<0.001	<0.001	<0.001	<0.001
Site III	<0.001	<0.001	<0.001	<0.001	<0.001	<0.001
Site IV	<0.001	<0.001	<0.001	<0.001	<0.001	<0.001

TABLE 6.4
Pb Concentrations (in mg L^{-1}) of Water from Different Study Sites

Study Sites	2013–2014			2014–2015		
	Rainy	Winter	Summer	Rainy	Winter	Summer
Site I	<0.001	<0.001	<0.001	<0.001	<0.001	<0.001
Site II	<0.001	<0.001	<0.001	<0.001	<0.001	<0.001
Site III	<0.001	<0.001	<0.001	<0.001	<0.001	<0.001
Site IV	<0.001	<0.001	<0.001	<0.001	<0.001	<0.001

TABLE 6.5
Cr Concentrations (in mg L⁻¹) of Water from Different Study Sites

Study Sites	2013–2014			2014–2015		
	Rainy	Winter	Summer	Rainy	Winter	Summer
Site I	<0.001	<0.001	<0.001	<0.001	<0.001	<0.001
Site II	<0.001	<0.001	<0.001	<0.001	<0.001	<0.001
Site III	<0.001	<0.001	<0.001	<0.001	<0.001	<0.001
Site IV	<0.001	<0.001	<0.001	<0.001	<0.001	<0.001

TABLE 6.6
Zn Concentrations (in mg L⁻¹) of Water from Different Study Sites

Study Sites	2013–2014			2014–2015		
	Rainy	Winter	Summer	Rainy	Winter	Summer
Site I	<0.02	<0.02	<0.02	<0.02	<0.02	<0.02
Site II	<0.02	<0.02	<0.02	<0.02	<0.02	<0.02
Site III	<0.02	<0.02	<0.02	<0.02	<0.02	<0.02
Site IV	<0.02	<0.02	<0.02	<0.02	<0.02	<0.02

Zinc (Zn)

Zn was present in a minute negligible amount of <0.02 mg L^{-1} for all the samples collected from the different sites, as shown in Table 6.6. In this study, all the values of Zn concentrations are found far below the limit and are not harmful to use for any purposes.

MACROPHYTES/WETLAND PLANT SPECIES COMPOSITION IN A RAMSAR SITE OF BIODIVERSITY HOTSPOT

From the phytosociological studies of the different sites, altogether, a total of 24 wetland macrophytes species belonging to 23 genera and 17 families were recorded. Of this, 10 species belonging to 8 genera and 8 families, 13 species belonging to 12 genera and 11 families, 12 species belonging to 11 genera and 9 families, and 21 species belonging to 20 genera and 15 families were reported from site I, II, III, and IV, respectively (Table 6.7).

The macrophyte species, namely, *Alternanthera philoxeroides* Griseb., *Arthraxon lanceolatus* (Roxb) Hochst., *A. pinnata* Lam., *C. demersum* Linn., *Cyrtococcum accrescens* Stafp., *E. crassipes* Linn., *Enhydra fluctuans* Lour., *E. ferox* Salisb., *H. verticillata* (Linn) Royle., *I. aquatica* Forsk., *L. minor* Linn., *N. nouchali* Burm. f., *N. cristatum* O. Kuntz, *Oenanthe javanica* (Bl) D.C., *P. stratiotes* Linn., *Polygonum glabrum* Willd., *P. crispus* Linn., *Rumex nepalensis* Spreng., *S. cucullata* Roxb.,

TABLE 6.7

Phytosociological Attributes of Macrophyte Species

Parameter	Site I	Site II	Site III	Site IV
Number of family	8	11	9	15
Number of genera	8	12	11	20
Number of species	10	13	12	21
Simpson index of dominance	0.37	0.27	0.42	0.12
Shannon–Wiener diversity index	1.31	1.68	1.37	2.37

TABLE 6.8

Sorenson's Similarity Index between Different Sites

	Site I	Site II	Site III	Site IV
Site I				
Site II	0.87			
Site III	0.73	0.72		
Site IV	0.65	0.71	0.61	

S. natans Hoffm., *Spirodela polyrhiza* (Linn) Schleid., *T. natans* Linn., *V. spiralis* Linn., and *Z. latifolia* (Griseb) stapf. were recorded.

A. philoxeroides Griseb., *A. pinnata* Lam., *E. crassipes* Linn., *L. minor* Linn., *P. stratiotes* Linn., *S. cucullata* Roxb., *S. natans* Hoffm., and *S. polyrhiza* (Linn) Schleid. were the plants with higher density al all the sites.

Table 6.8 shows the phytosociological attributes of macrophyte species in the study. Shannon–Wiener diversity index for macrophyte species was maximum at site IV, i.e., 2.37, and minimum at site I, i.e., 1.31. A reverse trend in the results was observed in case of the Simpson index of dominance. The Simpson index of dominance was maximum at site III, i.e., 0.42, and minimum at site IV, i.e., 0.12. Table 6.8 shows the present dominance and diversity. As the Simpson index of dominance values decrease, Shannon–Wiener diversity increases, which is quite appropriate.

SIMILARITY INDEX (SORENSON'S SIMILARITY INDEX) IN A RAMSAR SITE OF BIODIVERSITY HOTSPOT

Between site I and site II, the Sorenson's Similarity was found to be 0.87, between site I and site III, it was found to be 0.73; between site I and site IV, it was 0.65; between site II and site III, it was 0.72; between site II and site IV, it was 0.71; and between site III and site IV, it was found to be 0.61, as shown in Table 6.8.

From the phytosociological studies of the different sites, the macrophyte species common/similar to all the sites were *A. philoxeroides* Griseb., *A. pinnata* Lam., *E. crassipes* Linn., *L. minor* Linn., *P. stratiotes* Linn., *S. cucullata* Roxb., *S. natans* Hoffm., and *S. polyrhiza* (Linn) Schleid.

The uncommon/dissimilar species were *A. lanceolatus* (Roxb) Hochst., *C. demersum* Linn., *C. accrescens* Stafp., *E. fluctuans* Lour., *E. ferox* Salisb., *H. verticillata* (Linn) Royle., *I. aquatica* Forsk., *N. nouchali* Burm. f., *N. cristatum* O. Kuntz., *O. javanica* (Bl) D.C., *P. glabrum* Willd., *P. crispus* Linn., *R. nepalensis* Spreng., *T. natans* Linn., *V. spiralis* Linn., and *Z. latifolia* (Griseb) stapf., were recorded.

The majority of the species common to all sites belong to the families Amaranthaceae, Araceae, Azollaceae, Lemnaceae, Pontederiaceae, and Salviniaceae.

DOMINANCE OF FAMILIES AND DIVERSITY IN A RAMSAR SITE OF BIODIVERSITY HOTSPOT

A total of 17 families were recorded from all the study sites. Of these, 15 families were reported from site IV, followed by 11 families from site II, 9 families from site III, and 8 families from site I. Table 6.9 shows family-wise distribution of macrophyte species.

In site I, Lemnaceae and Salvinaceae were the dominant family, with two species each, followed by Amaranthaceae with one species, Araceae with one species, Azollaceae with one species, Ceratophyllaceae with one species, Hydrocharitaceae with one species, and Pontederiaceae with one species (Table 6.9).

Similarly, in site II, Lemnaceae and Salvinaceae were the dominant family, with two species each, followed by Amaranthaceae with one species, Araceae with one

TABLE 6.9
Family-Wise Distribution of Macrophyte Species

Sl. No.	Family	Site I	Site II	Site III	Site IV
1	Amaranthaceae	1	1	1	1
2	Apiaceae	–	–	–	1
3	Araceae	1	1	1	1
4	Asteraceae	–	1	–	–
5	Azollaceae	1	1	1	1
6	Ceratophyllaceae	1	1	–	1
7	Convolvulaceae	–	–	1	1
8	Hydrocharitaceae	1	1	–	2
9	Lemnaceae	2	2	2	2
10	Menyanthaceae	–	1	–	1
11	Nymphaeaceae	–	–	–	2
12	Poaceae	–	1	1	3
13	Polygonaceae	–	–	2	–
14	Pontederiaceae	1	1	1	1
15	Potamogetonaceae	–	–	–	–
16	Salviniaceae	2	2	2	2
17	Trapaceae	–	–	–	1

Note: –, absent.

species, Asteraceae with one species, Azollaceae with one species, Ceratophyllaceae with one species, Hydrocharitaceae with one species, Menyanthaceae with one species, Poaceae with one species, and Pontederiaceae with one species (Table 6.9).

Similarly, in site III, Lemnaceae and Salvinaceae, with one more family than site II, i.e., Polygonaceae, were the dominant families, with two species each, followed by Amaranthaceae with one species, Araceae with one species, Azollaceae with one species, Convolvulaceae with one species, Poaceae with one species, and Pontederiaceae with one species (Table 6.9).

Similarly, in site IV, Poaceae was dominant, with three species; followed by Lemnaceae, Hydrocharitaceae, Nymphaeaceae, and Salvinaceae family, with two species each; followed by Amaranthaceae with one species, Apiaceae with one species, Araceae with one species, Azollaceae with one species, Ceratophyllaceae with one species, Convolvulaceae with one species, Menyanthaceae with one species, Pontederiaceae with one species, Potamogetonaceae with one species, and Trapaceae with one species (Table 6.9).

Amaranthaceae, Araceae, Azollaceae, Lemnaceae Pontederiaceae, and Salviniaceae were the families common to all the study sites, i.e., site I, site II, site III, and site IV. Apiaceae and Trapaceae are restricted to a higher diversity site, i.e., site IV, whereas Asteraceae and Polygonaceae are restricted to low-diversity sites, i.e., site II and site III.

REFERENCES

Ahmad S, Reshi ZA, Shah MA, Rashid I, Ara R, Andrabi SMA. 2014. Phytoremediation Potential of *Phragmites australis* in Hokersar Wetland—A Ramsar site of Kashmir Himalaya. *Int J Phytoremediation* 16(12):1183–1191.

Ahmad S et al. 2016. Heavy metal accumulation in the leaves of *Potamogeton natans* and *Ceratophyllum demersum* in a Himalayan RAMSAR site: Management implications. *Wetlands Ecol Manage* 24:469–475.

Alhashemi A et al. 2011. Bioaccumulation of trace elements in trophic levels of wetland plants and waterfowl birds. *Biol Trace Elem Res* 142:500–516.

Alhashemi A et al. 2012. Bioaccumulation of trace elements in water, sediment, and six fish species from a freshwater wetland, Iran. *Microchem J* 104:1–6.

Baldantoni D, Alfani A, Di Tommasi P, Bartoli G, De Santo A. 2004. Assessment of macro and microelement accumulation capability of two aquatic plants. *Env. Poll* 130:149–156.

Baldantoni D, Maisto G, Bartoli G, Alfani A. 2005. Analyses of three native aquatic plant species to assess spatial gradients of lake trace element contamination. *Aquat. Bot* 83:48–60.

Bassi N, Dinesh Kumar M, Sharma A, Pardha-Saradhi P. 2014. Status of wetlands in India: A review of extent, benefits, threats ecosystem and management strategies. *Journal of Hydrology*: Regional Studies 2:1–19.

Chaterjee S et al. 2007. East Calcutta wetlands as a sink of industrial heavy metals: A PIXE study. *Int J PIXE* 17(3 & 4):129–142.

Chaterjee S et al. 2011. A study on the phytoaccumulation of waste elements in wetland plants of a Ramsar site in India. *Environ Monit Assess* 178:361–371.

Devi BN, Sharma BM. 2002. Life form analysis of the macrophytes of the Loktak Lake, Manipur, India. *Indian J Environ Ecoplan* 6(3):451–458.

Duplay J et al. 2013. Water quality in a protected natural wetland: El Kelbia sebkhet, Tunisia. *Int J Environ Stud* 70(1):33–48.

Fawzy MA, Badr NES, Khatib AE, Kassem AAE. 2012. Heavy metal biomonitoring and phytoremediation potentialities of aquatic macrophytes in River Nile. *Environ. Monit Assess* 184:1753–1771.

Feng XQ, Zhang GX, Jun Xu Y. 2013. Simulation of hydrological processes in the Zhalong wetland within a river basin, Northeast China. *Hydrol Earth System Sci* 17: 2797–2807.

Galal TM, Shehata HS. 2016. Growth and nutrients accumulation potentials of giant reed (*Arundo donax* L.) in different habitats in Egypt. *Int J Phytoremediation* 18(12):1221–1230.

García-García G, Nandini S, Sarma SS, Martínez-Jerónimo F, Jiménez-Contreras J. 2012. Impact of chromium and aluminium pollution on the diversity of zooplankton: A case study in the Chimaliapan wetland (Ramsar site) (Lerma basin, Mexico). *J Environ Sci Health Part A* 47(4):534–547.

Gardea-Torresdey JL, Peralta-Videa JR, Rosa GDL, Parson JG. 2005. Phytoremediation of heavy metals and study of the metal coordination by x-ray spectroscopy. *Coord Chem Rev* 17–18:1797–1810.

Garg A, Joshi B. 2015. Ecosystem sustenance of Upper Ganga Ramsar site through phytoremediation. *Geophytology* 45(2):175–180.

Gbogbo F, Otoo S. 2015. The concentrations of five heavy metals in components of an economically important urban coastal wetland in Ghana: Public health and phytoremediation implications. *Environ Monit Assess* 187:655.

Gosselink JG, Turner RE. 1978. The role of hydrology in freshwater wetland ecosystems. In *Freshwater Wetlands—Ecological Processes and Management Potential*, eds Good RE, Whigham DF, Simpson RL, 63–78. New York: Academic Press.

Grigoriadis N, Panagopoulos A, Meliadis I, Spyroglou G, Stathaki S. 2009. Habitat and hydrological–hydrochemical characteristics of the Agras wetland (Northern Greece). *Plant Biosyst* 143(1):162–172.

Loktak Development Authority (LDA). 2011. *Annual Administrative Report 2010–11*. Manipur, India: LDA.

Khatun A. 2016. Evaluation of metal contamination and phytoremediation potential of aquatic macrophytes of East Kolkata Wetlands, India. *Environ Health Toxicol* 31:7.

Kosygin L, Dhamendra H. 2009. Ecology and conservation of Loktak Lake, Manipur: An overview. In: *Wetlands of North East India Ecology, Aquatic Bioresources and Conservation*, ed. K Laishram, 1–20. New Delhi: Akansha Publishing House.

Kumar AB. 1999. Our vanishing wetlands. *Sci Rep* December:9–15.

Kwok CK et al. 2010. Ecotoxicological study on sediments of Mai Po marshes, Hong Kong using organisms and biomarkers. *Ecotoxicol Environ Saf* 73:541–549.

Maltby E, Acreman MC. 2011. Ecosystem services of wetlands: Pathfinder for a new paradigm. *Hydrol Sci J* 56(8):1341–1359.

Mbuligwe S, Kaseva ME, Kassenga GR. 2011. Applicability of engineered wetland systems for wastewater treatment in Tanzania—A review. *Open Environ Eng J* 4:18–31.

Mirza N et al. 2011. Ecological restoration of arsenic contaminated soil by *Arundo donax* L. *Ecol Eng* 37:1949–1956.

Mitsch WJ, Gosselink JG. 1986. *Wetlands*. New York: Van Nostrand Reinhold.

Nahlik AM, Mitsch WJ. 2006. Tropical treatment wetlands dominated by free-floating macrophytes for water quality improvement in Costa Rica. *Ecol Eng* 28:246–257.

Ningombam B, Bordoloi S. 2008. Loktak Lake, Manipur, India: A Congenial Habitat for the Amphibian Fauna. In *Proceeding of Taal 2008 (Jaipur, India)*, eds Sengupta M, Dalwani R, 519–524.

Pip E, Stepaniuk J. 1992. Cadmium, copper and lead in sediments and aquatic macrophytes in the Lower Nelson River system, Manitoba, Canada: I. Interspecific differences and macrophyte-sediment relations. *Archiv für Hydrobiologie* 124:337–355.

Pour SK, Monavari SM, Riazi B, Khorasani N. 2015. Caspian Rapid Assessment Method: A localized procedure for assessment of wetlands at southern fringe of the Caspian Sea. *Environ Monit Assess* 187:420.

Pourkhabbaz AR, Pourkhabbaz HR, Khazaei T, Behravesh S, Ebrahimpour M. 2011. Assessment of heavy metal accumulation in Anzali wetland, Iran, using a submerged aquatic plant, *Ceratophyllum demersum*. *Afr J Aquat Sci* 36(3):261–265.

Prasad SN, Ramachandra TV, Ahalya N, Sengupta T, Kumar A, Tiwari AK, Vijayan VS, Vijayan L. 2002. Conservation of wetlands of India—A review. *Trop Ecol* 43(1):173–186.

Rai PK. 2007a. Phytoremediation of Pb and Ni from industrial effluents using Lemna minor: An eco-sustainable approach. *Bull Biosci* 5(1):67–73.

Rai PK. 2007b. Wastewater management through biomass of *Azolla pinnata*: An ecosustainable approach. *Ambio* 36(5): 426–428.

Rai PK. 2008a. Heavy-metal pollution in aquatic ecosystems and its phytoremediation using wetland plants: An ecosustainable approach. *Int J Phytoremediation* 10(2):133–160.

Rai PK. 2008b. Phytoremediation of Hg and Cd from industrial effluents using an aquatic free floating macrophyte *Azolla pinnata*. *Int J Phytoremediation* 10(5):430–439.

Rai PK. 2008c. Mercury pollution from chlor-alkali industry in a tropical lake and its biomagnification in aquatic biota: Link between chemical pollution, biomarkers and human health concern. *Human Ecol Risk Assess Int J* 14:1318–1329.

Rai PK. 2009a. Heavy metal phytoremediation from aquatic ecosystems with special reference to macrophytes. *Cri Rev Env Sci Tec* 39(9):697–753.

Rai PK. 2009b. Heavy metals in water, sediments, and wetland plants in an aquatic ecosystem of tropical industrial region, India. *Environ Monit Assess* 158:433–457.

Rai PK. 2012. Assessment of multifaceted environmental issues and model development of an Indo-Burma hot spot region. *Environ Monit Assess* 184:113–131.

Rai PK. 2013. *Plant Invasion Ecology: Impacts and Sustainable Management*. New York: Nova science Publisher, pp. 160.

Rai PK. 2015a. Paradigm of plant invasion: Multifaceted review on sustainable management. *Environ Monit Assess* 187:759–785.

Rai PK. 2015b. What makes the plant invasion possible? Paradigm of invasion mechanisms, theories and attribute. *Environ Skeptics Critics* 4(2):36–66.

Rai PK. 2015c. Concept of plant invasion ecology as prime factor for biodiversity crisis: Introductory review. *Int Res J Environ Sci* 4(5):85–90.

Rai PK. 2015d. *Environmental Issues and Sustainable Development of North East India*. Germany: Lambert Academic Publisher, pp. 308.

Rai PK. 2016. Biodiversity of roadside plants and their response to air pollution in an Indo-Burma hotspot region: Implications for urban ecosystem restoration. *J Asia Pac Biodivers* 9:47–55.

Rai PK, Chutia B. 2016. Biomagnetic monitoring through *Lantana* leaves in an Indo-Burma hot spot region. *Environ Skeptics Critics* 5(1):1–11.

Rai PK, Singh MM. 2014. Wetland Resources of Loktak Lake in of Manipur, India: A Review. *Sci and Technol J* 2(1):98–103.

Rai PK, Singh MM. 2015. *Lantana camara* invasion in urban forests of an Indo-Burma hotspot region and its ecosustainable management implication through biomonitoring of particulate matter. *J Asia Pac Biodivers* 8:375–381.

Rai PK, Singh MM. 2016. *Eichhornia crassipes* as a potential phytoremediation agent and an important bioresource for Asia Pacific region *Environ Skeptics Critics* 5(1):12–19.

Rai PK, Tripathi BD. 2009. Comparative assessment of *Azolla pinnata* and *Vallisneria spiralis* in Hg removal from G.B. Pant Sagar of Singrauli Industrial region, India. *Environ Monit Assess* 148:75–84.

Rai SC, Raleng A. 2011. Ecological studies of wetland ecosystem in Manipur valley from management perspectives. In *Ecosystems Biodiversity*, eds Grillo O, Venora G, 233–248. Rijeka, Croatia: InTech.

Ramsar Convention Secretariat. 2013. *The Ramsar Convention Manual: A Guide to the Convention on Wetlands (Ramsar, Iran, 1971)*. 6th ed. Gland, Switzerland: Ramsar Convention Secretariat.

Rodríguez-Iruretagoiena A et al. 2016. Uptake and distribution of trace elements in dominant mangrove plants of the Indian Sundarban wetland. *Bull Environ Contam Toxicol* 97:721–727.

Roy RD. 1992. *Case Study of LoktakLake of Manipur. Wetlands of India*, 37–70. India: Ashish Publishing House.

SAC (Space Applications Centre, ISRO, India). 2011. Information Brochure National Wetland Inventory & Assessment. Sponsored by Ministry of Environment and Forests, Govt. of India, p. 32. Available at http://www.moef.nic.in/downloads/public-information/NWIA _National_brochure.pdf

Sanjit L, Bhatt D, Sharma RK. 2005. Habitat heterogeneity of the Loktak Lake, Manipur. *Curr Sci* 88(7):1027–1028.

Santra SC. 2001. *Environmental Science*. Kolkata, India: New Central Book Agency (P) Ltd.

Sereyrath L, Irvine KN, Murphy TP, Wilson K. 2016. Sediment associated metals levels along the sewer–natural treatment wetland continuum, Phnom Penh, Cambodia. *Urban Water J* 13(8):819–829.

Sharma S, Sujatha D. 2016. Characterization of the water chemistry, sediment ^{13}C and ^{18}O compositions of Kolleru Lake—A Ramsar wetland in Andhra Pradesh, India. *Environ Monit Assess* 188:409.

Singh MM, Rai PK. 2014. A study of depleting water quality and wildlife resources of the Loktak Lake, Manipur, India. In *Issues and Trends of Wildlife Conservation in Northeast India*, 96–104. Mizo Academy of Sciences.

Singh MM, Rai PK. 2016. Microcosm investigation of Fe (iron) removal using macrophytes of Ramsar lake: A phytoremediation approach. *International Journal of phytoremediation* 18(12):1231–1236.

Singh NS, Devi CB, Sudarshan M, Meetei NS, Singh TB, Singh NR. 2013. Influence of Nambul River on the quality of fresh water in Loktak Lake. *International Journal of Water Resources and Environmental Engineering* 5(6):321–327.

Sood A, Uniyal PP, Prasanna R, Ahluwalia AS. 2012. Phytoremediation Potential of Aquatic Macrophyte, Azolla. *Ambio* 41:122–137.

Taggart MA et al. 2009. Arsenic rich iron plaque on macrophyte roots—An ecotoxicological risk? *Environ Pollut* 157:946–954.

Trivedy PK, Goel PK. 1986. *Chemical and Biological Methods for Water Pollution Studies, Series in Methodology*. Karad, India: Environmental Publications, 220 pp.

Tombi H, Shyamananda RK. 1994. *Loktak*. New Delhi: World Wide Fund.

Uj B et al. 2016. Wetland habitats of the Kis-Sárrét 1860–2008 (Körös-Maros National Park, Hungary). *J Maps* 12(2):211–221.

Verma M, Negandhi D. 2011. Valuing ecosystem services of wetlands—A tool for effective policy formulation and poverty alleviation. *Hydrol Sci J* 56(8):1622–1639.

7 Global Wetland Plants in Metal/Metalloid Phytoremediation
Microcosm and Field Results

PHYTOREMEDIATION OF EMERGING CONTAMINANTS IN A RAMSAR SITE: FIELD INVESTIGATION OF METALS IN WETLAND PLANTS OF GLOBAL BIODIVERSITY HOTSPOT

Four macrophytes/wetland plant species samples were collected from all the four sites in the winter season from a global biodiversity hotspot as mentioned in a previous chapter (2014–2015). Only these macrophytes were investigated for the metal concentration inside the biomass. The results for the heavy metal concentrations in the collected macrophyte/wetland plant species are as follows.

IRON (FE)

The highest Fe concentration measured was 28.29 mg/kg, in *Pistia stratiotes* at Site II, which is higher than the permissible limit set by the World Health Organization, and the lowest was 1.68 mg/kg, in *Salvinia cucullata* at Site I, as shown in Figure 7.1 and Table 7.1. *S. cucullata* has the lowest Fe concentration compared with the other plant species, that is, *Eichhornia crassipes*, *Lemna minor*, and *P. stratiotes*. The results for the Fe concentrations of the plant species from different sites wise were grouped together with the Fe concentration of the water from different sites and correlated to find their correlation coefficient.

MERCURY (HG)

The highest Hg concentration measured was <0.0069 mg/kg, in *S. cucullata* at Site I, and the lowest was <0.0017 mg/kg, in *E. crassipes* at Site I, as shown in Table 7.2. *E. crassipes* has the lowest Hg concentration compared with the other macrophyte species, that is, *L. minor*, *P. stratiotes*, and *S. cucullata*.

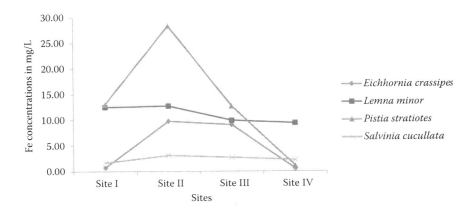

FIGURE 7.1   Variations in Fe concentrations (in mg/kg) of plants from different study sites.

TABLE 7.1
Fe Concentrations (in mg/kg) of Plants from Different Study Sites

Name of the Plant	Site I	Site II	Site III	Site IV
Eichhornia crassipes	0.72	9.77	9.07	0.53
Lemna minor	12.52	12.74	9.92	9.41
Pistia stratiotes	13.01	28.29	12.68	1.09
Salvinia cucullata	1.68	3.14	2.72	2.25

TABLE 7.2
**Variations in Hg Concentration (in mg/kg) of Plants
from Different Study Sites**

Name of Plant	Site I	Site II	Site III	Site IV
Eichhornia crassipes	<0.00173	<0.00328	<0.00249	<0.00178
Lemna minor	<0.00318	<0.00309	<0.00228	<0.00318
Pistia stratiotes	<0.00356	<0.00417	<0.00247	<0.00389
Salvinia cucullata	<0.00698	<0.00314	<0.00368	<0.00234

Cadmium (Cd)

Similarly as with Hg concentration, the highest Cd concentration measured was <0.0069 mg/kg, in *S. cucullata* at Site I, and the lowest was 0.0017 mg/kg, in *E. crassipes* at Site I, as shown in Table 7.3. *E. crassipes* has the lowest Cd concentration compared with the other macrophyte species, that is, *L. minor*, *P. stratiotes*, and *S. cucullata*.

TABLE 7.3
Cd Concentrations (in mg/kg) of Plants from Different Study Sites

Name of Plant	Site I	Site II	Site III	Site IV
Eichhornia crassipes	<0.00173	<0.00328	<0.00249	<0.00178
Lemna minor	<0.00318	<0.00309	<0.00228	<0.00318
Pistia stratiotes	<0.00356	<0.00417	<0.00247	<0.00389
Salvinia cucullata	<0.00698	<0.00314	<0.00368	<0.00234

ARSENIC (As)

Similarly as with Hg and Cd concentrations, the highest As concentration measured was <0.0069 mg/kg, in *S. cucullata* at Site I, and the lowest was 0.0017 mg/kg, in *E. crassipes* at Site I, as shown in Table 7.4. *E. crassipes* has the lowest As concentration compared with the other macrophyte species, that is, *L. minor*, *P. stratiotes*, and *S. cucullata*.

LEAD (Pb)

Similarly as with Hg, Cd, and As concentrations, the highest Pb concentration measured was <0.0069 mg/kg, in *S. cucullata* at Site I, and the lowest was 0.0017 mg/kg, in *E. crassipes* at Site I, as shown in Table 7.5. *E. crassipes* has the lowest Pb concentration compared with the other macrophyte species, that is, *L. minor*, *P. stratiotes*, and *S. cucullata*.

TABLE 7.4
As Concentrations (in mg/kg) of Plants from Different Study Sites

Name of Plant	Site I	Site II	Site III	Site IV
Eichhornia crassipes	<0.00173	<0.00328	<0.00249	<0.00178
Lemna minor	<0.00318	<0.00309	<0.00228	<0.00318
Pistia stratiotes	<0.00356	<0.00417	<0.00247	<0.00389
Salvinia cucullata	<0.00698	<0.00314	<0.00368	<0.00234

TABLE 7.5
Pb Concentrations (in mg/kg) of Plants from Different Study Sites

Name of Plant	Site I	Site II	Site III	Site IV
Eichhornia crassipes	<0.00173	<0.00328	<0.00249	<0.00178
Lemna minor	<0.00318	<0.00309	<0.00228	<0.00318
Pistia stratiotes	<0.00356	<0.00417	<0.00247	<0.00389
Salvinia cucullata	<0.00698	<0.00314	<0.00368	<0.00234

TABLE 7.6
Cr Concentrations (in mg/kg) of Plants from Different Study Sites

Name of Plant	Site I	Site II	Site III	Site IV
Eichhornia crassipes	<0.00173	<0.00328	<0.00249	<0.00178
Lemna minor	<0.00318	<0.00309	<0.00228	<0.00318
Pistia stratiotes	<0.00356	<0.00417	<0.00247	<0.00389
Salvinia cucullata	<0.00698	<0.00314	<0.00368	<0.00234

TABLE 7.7
Zn Concentrations (in mg/kg) of Plants from Different Study Sites

Name of Plants	Site I	Site II	Site III	Site IV
Eichhornia crassipes	<0.0345	<0.0656	<0.04983	<0.03552
Lemna minor	<0.06355	<0.06184	<0.0456	<0.06355
Pistia stratiotes	<0.07128	<0.08344	<0.04943	<0.07788
Salvinia cucullata	<0.13966	<0.06279	<0.07356	<0.04688

CHROMIUM (CR)

Similarly as with Hg, Cd, As, and Pb concentrations, the highest Cr concentration measured was <0.0069 mg/kg, in *S. cucullata* at Site I, and the lowest was 0.0017 mg/kg, in *E. crassipes* at Site I, as shown in Table 7.6. *E. crassipes* has the lowest Cr concentration compared with the other macrophyte species, that is, *L. minor*, *P. stratiotes*, and *S. cucullata*.

ZINC (ZN)

The highest Zn concentration measured was <0.13 mg/kg, in *S. cucullata* at Site I, and the lowest was 0.03 mg/kg, in *E. crassipes* at Site I, as shown in Table 7.7. *E. crassipes* has the lowest Cr concentration compared with the other macrophyte species, that is, *L. minor*, *P. stratiotes*, and *S. cucullata*.

PHYTOREMEDIATION OF EMERGING CONTAMINANTS (HEAVY METALS) IN RAMSAR SITE OF GLOBAL BIODIVERSITY HOTSPOT: MICROCOSM INVESTIGATION

Phytoremediation experiments were conducted using selected macrophyte species that were available at all the sites, that is, Sites I, II, III, and IV, as per the study of the phytosociology of the sites. The macrophytes that were used in the phytoremediation experiments are *E. crassipes*, *L. minor*, *P. stratiotes*, and *S. cucullata*. As per the results of the heavy metal analysis of water from the different sites, we conducted the experiment using Fe only as it was the metal found in the highest amount, that is, above the permissible limit. The results are given in Tables 7.8 through 7.10.

Fe removal in water under varying ranges of metal concentration, that is, 1, 3, and 5 mg/L, was maximum in E. *crassipes*-treated water followed by P. *stratiotes*-treated water, L. *minor*-treated water, and S. *cucullata*-treated water for intervals of 4, 8, and 18 days (Singh and Rai, 2016). Student's *t* test revealed that the decrease in Fe concentrations was significant (at $p < .01$) when compared with the control.

TABLE 7.8

Percentage (%) Removal of Fe by Selected Plant Species of Different Concentrations in 4 Days

Concentration (mg/L)	Eichhornia Crassipes	Lemna Minor	Pistia Stratiotes	Salvinia Cucullata
1	68	25	27	39
3	40.3	21	36	30
5	45.4	34.4	36.8	26

TABLE 7.9

Percentage (%) Removal of Fe by Selected Plant Species of Different Concentrations in 8 Days

Concentration (mg/L)	Eichhornia Crassipes	Lemna Minor	Pistia Stratiotes	Salvinia Cucullata
1	81	71	78	76
3	55	59.3	61.3	40.3
5	76.6	48	45.2	58

TABLE 7.10

Percentage (%) Removal of Fe by Selected Plant Species of Different Concentrations in 12 Days

Concentration (mg/L)	Eichhornia Crassipes	Lemna Minor	Pistia Stratiotes	Salvinia Cucullata
1	89	87	87	81
3	81.3	62.3	75.7	59.3
5	73.2	69.6	63.4	61.4

In the 4-day experiment, *E. crassipes*-treated water showed maximum of 68% removal from 1 mg/L, 40.3% removal from 3 mg/L, and 45.4% removal from 5 mg/L; *L. minor*-treated water showed minimum of 25% removal from 1 mg/L and 21% removal from 3 mg/L; and *S. cucullata*-treated water showed 26% removal from 5 mg/L as shown in Figure 7.2 (Singh and Rai, 2016).

In the 8-day experiment, *E. crassipes*-treated water showed maximum of 81% removal from 1 mg/L, 55% removal from 3 mg/L, and 76.6% removal from 5 mg/L; *L. minor*-treated water showed minimum of 71% removal from 1 mg/L; *S. cucullata*-treated water showed 40.3% removal from 3 mg/L; and *P. stratiotes*-treated water showed 45.2% removal from 5 mg/L as shown in Figure 7.3.

In the 12-day experiment, *E. crassipes*-treated water showed maximum of 89% removal from 1 mg/L, 81.3% removal from 3 mg/L, and 73.2% removal from 5 mg/L

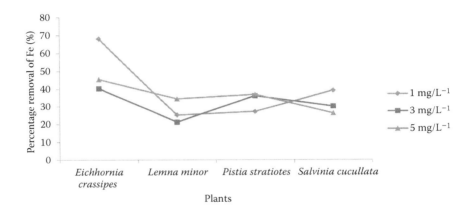

FIGURE 7.2 Graph showing percentage removal of Fe (%) by selected plant species of different concentrations in 4 days.

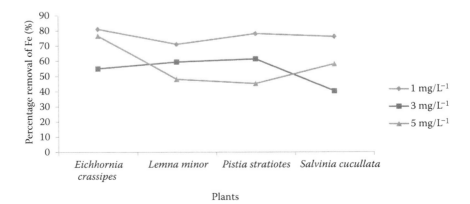

FIGURE 7.3 Graph showing percentage removal of Fe (%) by selected plant species of different concentrations in 8 days.

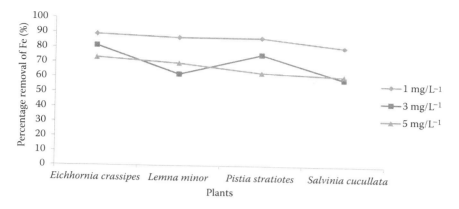

FIGURE 7.4 Graph showing percentage removal of Fe (%) by selected plant species of different concentrations in 12 days.

and *S. cucullata*-treated water showed minimum of 81% removal from 1 mg/L, 59.3% removal from 3 mg/L, and 61.4% removal from 5 mg/L as shown in Figure 7.4.

Although I confined my quest for the phytoremediation potential of four macrophytes pertaining to Fe only in the context of the present status of Loktak Lake pollution (Figure 7.5), future microcosm studies will also investigate the phytoremediation potential of other observed macrophytes for different ranges of heavy metals.

FIGURE 7.5 (a, b) Microcosm case study on wetland plants of floating nature *E. crassipes*, *L. minor*, *S. cucullata* (top), and *P. stratiotes* (bottom) during our phytoremediation experiment (Singh and Rai, 2016) on the removal of Fe.

CONCLUDING REMARKS ON INVESTIGATION ON RAMSAR WETLAND SITE OF GLOBAL BIODIVERSITY HOTSPOT

The result of the present study reveals that the lake is under severe stress mainly due to human interventions such as the polluted rivers draining in the lake, which originate from municipal areas, weed infestation, pollution, encroachment, overexploitation of resources, and siltation, thereby causing flooding of agricultural fields and villages, decrease in flora and fauna production, and loss of biodiversity.

The temperature of water is a function of seasonal ambient air temperatures. The temperature is one of the important factors in an aquatic environment since it regulates physicochemical as well as biological activities. Water in the temperature range of 7–11°C has a pleasant taste and is refreshing. At higher temperature, with less dissolved gases, water becomes tasteless and cannot even quench thirst (Rai, 2010a–d).

There was an increase in temperature with increasing degree of disturbance with some exception. The discharge of pollutants from the surrounding field area through runoff along with the rivers that drain into the lake may be attributed to high temperature during the rainy season, as energy is released during decomposition of waste present in the water. The decomposition of wetland macrophytes of the lake may also be a factor in the increase in water temperature. The temperature of water samples may not be that important for normal uses, but in polluted water, it has profound effect on the dissolved oxygen (DO), biological oxygen demand (BOD), and aquatic life (Kangabam et al., 2017). Similar trends were found by Khan et al. (2012), Singh et al. (2010), Umavathi and Logankumar (2010), and Zafar and Sultana (2008).

A positive and significant correlation was obtained for temperature with acidity and turbidity in all the study sites. On the contrary, a negative and significant correlation was obtained between transparency and alkalinity in all the study sites (Appendices I through IV).

In other words, the pH value of water is a measure of its acidity or alkalinity and is a very important indicator of its quality. pH influences the growth of plants and soil organism; therefore, it affects, to a great extent, the suitability of water for irrigation. The pH value of water is controlled by the amounts of bicarbonates, carbonates, and dissolved carbon dioxide.

There was a decrease in pH during the rainy season for all the sites. The pH is also below the permissible limit set by various agencies. In the rainy season, as the water is diluted with rainwater, the pH should be increasing, but on the contrary, the result shows a decreased pH value. This may be due to the late raining condition in the year along with the early sampling. It may also be due to the location of the sampling point, where the water is contaminated with the polluting material nearby.

The decreased pH may be due to several various factors such as interaction of water with suspended matter, polluting material from the runoff and polluted rivers that drain into the lakes, and decaying matter and other chemicals that were applied in the nearby field for agriculture and other purposes. Significant changes in pH occur due to the discharge of agricultural and domestic waste (Khan et al., 2012). The minerals present in the water also affect the pH. Most chemical and biochemical reactions are influenced by the pH of water. The toxicity of water increases with increasing acidic content in the water and is well cited elsewhere (Rai, 2010a–d).

A positive and significant correlation was obtained for pH with hardness. On the contrary, a negative correlation was obtained between total solids (TS) in all the study sites (Tables A.1 through A.4).

Transparency of water is one of the important water quality parameters. It indicates the clarity of water depending upon the presence of material in the aquatic bodies. Transparent waters allow more light penetration, which has far-reaching effects on all aquatic organisms, including their development, distribution, and behavior (Rai, 2010a–d). Transparent water is considered as good water because it allows light to penetrate inside where the plants growing inside will perform photosynthesis.

Transparency varies with the number of minute impurities present in the water bodies. Higher transparency was observed during winter and summer due to the absence of rain, runoff, and floodwater as well as the gradual settling of suspended particles (Khan and Chowdhury, 1994).

The high value of transparency in winter is due to the better settlement of solid particles in the waterbed, which helps the light to penetrate the water, while transparency is low in the rainy season due to the abundance of organisms, including macrophytes, on the surface of the water and solid materials in the water. Kosygin and Dhamendra (2009) noted that the transparency values of this lake range from 0.51 to 2.98 m.

A positive and significant correlation was obtained for transparency with alkalinity and chloride content in all sites except for Site II. On the contrary, a negative and significant correlation was obtained between transparency and acidity at all the study sites (Tables A.1 through A.4).

TS are solid substances present in water bodies whether dissolved or nondissolved. They disturb the penetration of light in water bodies. The higher the TS, the lower will be the survival of the aquatic organisms. TS are determined as the residue left after evaporation of the unfiltered sample. A higher value of TS indicates a more turbid water due to the presence of silt and organic matter.

The higher TS in the rainy season is due to the several various factors such as suspended matter, polluting material from the runoff and polluted rivers that drain into the lakes, decaying matter, and other organic material. The muddy agitated flowing mixed rainwater from the river that afterwards drains into the lake mixed with the lake water, affecting the clarity of water and increasing the TS.

A positive and significant correlation was obtained for TS with BOD in Site III. On the contrary, a negative and significant correlation was obtained between hardness, nitrate content, and phosphate content at all the study sites (Tables A.1 through A.4),

DO is one of the most important parameters of water quality that reflect the various physical and biological processes in water. All living organisms in water bodies are dependent upon DO in one form or another for carrying out the metabolic processes that produce energy for growth and reproduction. Low DO concentrations (<3 mg/L) in freshwater aquatic systems indicate a high pollution level of the waters and cause negative effects on life in this system (Rai, 2011).

DO content was low in summer because of enhanced utilization by microorganisms in the fast decomposition of organic matter, which indicates a high pollution load in the water. The fluctuations in oxygen content depend on factors such as temperature, decompositional activities, photosynthesis, and the level of aeration. The reoxygenation of water during a monsoon might be occurring due to circulation and mixing by inflow after monsoon rains. In summer, the DO depletion was due to high temperature. Also, low DO may be due to the luxuriant growth of algae and aquatic macrophytes or wetland plants, resulting in a higher photosynthetic rate as a result of the increased temperature (Nybakken, 1997). The oxygen value is indicative of pollution in water and has an inverse relationship with water temperature. Similar trends of DO in freshwater lakes were also observed by Khan et al. (2012) and Rai (2011).

A positive and significant correlation was obtained for DO with acidity. On the contrary, a negative and significant correlation was obtained between turbidity at all the study sites (Tables A.1 through A.4).

BOD is the most important parameter of water quality. BOD refers to the quantity of oxygen required by bacteria and other microorganisms for biochemical degradation and transformation of organic matter under aerobic conditions (Manivaskasam, 1986). The BOD can be used as a measure of the amount of organic materials present in an aquatic solution that support the growth of microorganisms. The BOD determines the strength of pollution in natural waters. The BOD test is applied for freshwater sources (rivers, lakes), wastewater (domestic, industrial), and marine water (estuaries, coastal water).

The high BOD of Site I is due to the polluted Nambul River and the domestic waste from the local areas, including waste from the residents around the lake itself. It also may be due to the high concentration of organic waste discharged into the water, leading to a high rate of decomposition and resulting in higher consumption of oxygen by the microorganisms. The enormous growth of aquatic macrophytes may lead to the high BOD of the site. Hacioglu and Dulger (2009), Naik (2005), and Rai (2011) reported similar trends.

A positive and significant correlation was obtained for BOD with chloride content in Site I. On the contrary, a negative and significant correlation was obtained between BOD and phosphate content (Tables A.1 through A.4).

The acidity of the water is its capacity to neutralize a strong base and is mostly due to the presence of strong mineral acids, weak acids (carbonic and acetic acids), and salts of strong acids and weak bases (e.g., ferrous sulfate, aluminum sulfate). These salts on hydrolysis produce strong acids and metal hydroxides, which are sparingly soluble, thus producing acidity. Determination of acidity is significant as it causes corrosion and influences chemical and biochemical reactions (Kulkarni and Shrivastava, 2000).

The high acidity in the rainy season may be due to the presence of a high organic load received from runoff, rivers, and decomposed macrophyte material in the water. Acidic water is less buffered and less productive because a sufficient amount of bicarbonates is not dissolved to give CO_2 for a high rate of photosynthesis. Lowering of the pH of water is a result of the decomposition of organic matter and final release of CO_2 (Rai, 2011).

The acidity of water refers to the amount of acids and bases present in it. The adverse effects of most acids appear below pH 5.0. There has been no particular limit for acidity, and it can be expressed in terms of $CaCO_3$. Highly acidic water could be dangerous and must be avoided. Acidity is not desirable in municipal water systems because it tends to increase corrosion.

A positive and significant correlation was obtained for acidity with temperature, DO, and turbidity in all the sites except for Site IV. On the contrary, a negative and significant correlation was obtained between phosphate content and acidity (Tables A.1 through A.4).

The acid-neutralizing capacity of water is known as alkalinity. The constituents of alkalinity in a natural system mainly include carbonates, bicarbonates, and hydroxides. These constituents result from dissolution of mineral substances in the soil and atmosphere (Mittal and Verma, 1997).

The presence of carbonates and bicarbonates may be the reason for high alkalinity in the winter season. It may originate from the microbial decomposition of organic matters. Higher values of alkalinity in winter months may be due to the liberation of carbon dioxide during decomposition, which reacts with water to form bicarbonate and also, during winter volume of water is relatively low which further add to higher alkalinity values. Water with low alkalinity and with pH in the range of 6.3–7.3 is low in production and supports phytoplankton, which have low acid and low alkaline adaptability.

A positive and significant correlation was obtained for alkalinity with nitrate and phosphate contents at all the sites except for Site IV. On the contrary, a negative and significant correlation was obtained with temperature (Tables A.1 through A.4).

Chloride content is one of the important water parameters, and chloride is found in nature in the form of salts of sodium, potassium, and calcium. Chloride is one of the most stable components of water, being unaffected by most physicochemical and/or biological processes. The major sources of chloride include natural mineral deposits, seawater intrusion, and agricultural and surface runoffs. Chlorides occur naturally in all types of waters. One of the most important sources of chlorides in water is the discharge of municipal and agricultural waste. Humans and other animals excrete very high quantities of chlorides together with nitrogenous compounds (Rai, 2011).

The high amount of chlorides in the winter season is due to the mixing of discharge of municipal and domestic waste from the polluted river, which then drains into the lake, agricultural waste, and surface runoff. Chlorides in water are indicators of a large amount of nonpoint source pollution by pesticides, grease, oil, metals, and other toxic materials (Khare and Jadhav, 2008; Rai, 2011). Higher values of chloride consumption are hazardous to humans and create health problems (Kataria and Iqbal, 1995). The chloride in water is under the permissible value set by different agencies. A similar trend of results was observed by Rai (2011), Singh et al. (2010), and Zafar and Sultana (2008).

A positive and significant correlation was obtained for chloride content with nitrate and phosphate contents at all the sites except for Sites I and IV. On the contrary, a negative and significant correlation was obtained between hardness and chloride content at site IV (Tables A.1 through A.4).

Hardness is the property of water that prevents lather formation with soap and increases the boiling point of water. Water hardness is a traditional measure of the capacity of water to precipitate soap. Calcium and magnesium are the principal cations causing hardness. However, other cations such as strontium, iron, and manganese also contribute to hardness. The anions responsible for hardness are mainly bicarbonate, carbonate, sulfate, chloride, nitrate, silicates, and so on (Rai, 2011).

A high value of hardness indicates a high concentration of calcium and magnesium in water bodies. In freshwater, the principal hardness-causing ions are calcium and magnesium. which precipitate soap. Other polyvalent cations may also precipitate soap, but often are in a complex form, frequently with organic constituents, and their role in water hardness may be minimal and difficult to define. Similar trends were found by Kumar (2000), Rai (2011), and Palharya et al. (1993).

A positive and significant correlation was obtained for hardness with pH in all the sites. On the contrary, a negative and significant correlation was obtained between TS and hardness (Tables A.1 through A.4).

Turbidity is an expression of the optical property of a water sample containing insoluble substances that cause light to be scattered rather than transmitted in straight lines. In most waters, turbidity is due to colloidal and extremely fine dispersions. Suspended matter such as clay, slit, finely divided organic and inorganic matter, plankton. and other microscopic organisms also contribute to turbidity (Rai, 2011).

Turbidity can be measured by its effect on the scattering light, through a method termed *nephelometry*. A turbidimeter can be used for samples with moderate turbidity and a nephelometer for samples with low turbidity. The higher the intensity of the scattered light, the higher the turbidity value.

Higher turbidity values were obtained during the summer season. This could be attributed to raining in the summer season when pollution load is discharged into the lake through the rivers that drain into it and also to surface runoff from the adjoining agricultural fields and fish farms, which makes the water more turbid.

High turbidity is due to silt, clay, and other suspended particles that contribute to turbidity during the summer and rainy seasons, whereas settlement of silt and clay in winter results in low turbidity. Decomposition is also high in summer due to high microbial activities and the production of organic microbial waste.

A positive and significant correlation was obtained for turbidity with temperature at all the sites. On the contrary, a negative and significant correlation was obtained between DO and turbidity (Tables A.1 through A.4).

Nitrate is the stable form of oxidized nitrogen, but it can be reduced by microbial action to nitrite. In the process of nitrification, nitrogen changes to a nitrate ion. Therefore, all sources of nitrogen (including organic nitrogen, ammonia, and fertilizers) should be considered as potential source of nitrates. The maximum limit of nitrate in drinking water is 45 mg/L for humans and 100 mg/L for livestock (Rai, 2011).

The high nitrate concentration in the summer season may be due to the higher decomposition rate of macrophytes and other materials in the water. Agricultural chemical fertilizers from the adjoining areas may also help in increasing the nitrate content of the water.

A high concentration of nitrate in drinking water is toxic (Rai, 2011). The concentration of nitrate ions in public water supplies is very important because a high nitrate concentration causes methemoglobinemia in children (Rai, 2011). The presence of nitrate indicates the organic pollution of water and not only causes cyanosis among infants when present in a considerable quantity but also has been reported to cause gastric cancer when present in a high quantity (Rai, 2011).

A positive and significant correlation was obtained for nitrate concentration with pH in all the sites. On the contrary, a negative and significant correlation was obtained between TS and nitrogen concentration (Tables A.1 through A.4).

Phosphate is usually derived from the leaching of phosphorus-rich bedrock and additionally from human waste, synthetic detergents, and industrial and agriculture waste. It is an important micronutrient for macrophytes and plays an important role in the growth of macrophytes, including phytoplankton. Phosphate helps in eutrophication.

The high phosphate content in summer may be due to the high decomposition of the macrophytes, thus leading to the liberation of phosphate in the water. The agricultural waste and agricultural chemical fertilizers with phosphate content used in the nearby field may also affect the water. Banerjee and Gupta (2010) and Umavathi and Longakumar (2010) reported similar trends. Consumption of drinking water with high phosphate content may cause osteoporosis and kidney damage.

A positive and significant correlation was obtained for acidity with temperature, DO, and turbidity in all the sites except for Site IV. On the contrary, a negative and significant correlation was obtained between acidity, BOD, and TS (Tables A.1 through A.4).

HEAVY METAL ANALYSIS IN WATER AND WETLAND PLANTS AT RAMSAR SITE OF BIODIVERSITY HOTSPOT

Of all the heavy metals tested, that is, iron (Fe), mercury (Hg), cadmium (Cd), arsenic (As), lead (Pb), chromium (Cr), and zinc (Zn), Fe is the only metal found in a high amount in water as well as in the plant samples. The other metals were found in negligible amounts or in other words well below the permissible limit prescribed by scientific agencies.

Fe was high in water samples due to the pollution caused by the draining rivers and the domestic waste from the surroundings. Fe in excess of 0.3 mg/L causes staining of clothes and utensils. The water is also not suitable for processing of food, beverages, ice, dyeing, bleaching, and so on. The limit on Fe in water is based on aesthetic and taste considerations rather than its physiological effects (Trivedy and Goel, 1986). Kakati and Bhattacharya (1990) studied the water quality of various surface water sources of greater Guwahati and found that Fe content ranges from 0.112 to 12.8 mg/L. High doses of Fe are known to cause hemorrhagic necrosis, sloughing of mucosa areas in the stomach, and tissue damage to a variety of organs by catalyzing the conversion of H_2O_2 to free radical ions that attack cell membranes and proteins, break the deoxyribonucleic acid (DNA)) double strands, and cause oncogene activation (Gurzau et al., 2003; Rai, 2011).

Iron is one of the elements vital to humans and other forms of life. Past studies have documented the Fe phytoremediation ability of obnoxious free-floating macrophytes from nutrient-rich wastewaters (Tripathi and Upadhyay, 2003; Sooknah and Wilkie, 2004; Jayaweera et al., 2008). *P. stratiotes* accumulate a high amount of Fe from the lake in site II. The accumulation of heavy metals such as Fe, Zn, Cu, Cr, and Cd did not cause any toxic effect on *P. stratiotes*, which qualifies the plant to be used for the phytoremediation of wastewater for heavy metals on a large scale (Singh and Rai, 2016). *P. stratiotes* has been considered a promising plant for the remediation of contaminated waters (Maine et al., 2001; Singh and Rai, 2016). Odjegba and Fasidi (2004) revealed the accumulation capacity and resistance of *P. stratiotes* against trace elements and concluded that it can tolerate heavy metals and the growth responses of the plant varied inversely with the increase in metal concentration.

PHYTOSOCIOLOGY OF WETLAND PLANTS: CONCLUDING REMARKS FOR A RAMSAR SITE OF BIODIVERSITY HOTSPOT

The physicochemical nature of environment and biological peculiarities of the macrophytes themselves play a significant role in the pattern of distribution of macrophytes. *Alternanthera philoxeroides* Griseb., *Azolla pinnata* Lam., *E. crassipes* Linn., *L. minor* Linn., *P. stratiotes* Linn., *S. cucullata* Roxb., *Salvinia natans* Hoffm., and *Spirodela polyrhiza* (Linn.) Schleid. were the plants with higher density in all the sites. This shows the resisting capacity and tolerance of these macrophytes with the variation in the degree of pollution and disturbance.

Many scientists and researchers globally are now using the diversity indices for various species of any given water body to find out the degree of pollution of that water body because the level of pollution is always equal to the loss of species biodiversity (Rai, 2011).

It is believed that pollution reduces biotic diversity (Rai, 2008; 2011). There were successive decreases in the number of species with increase in the degree of disturbance and pollution caused by anthropogenic activities. Due to this, Sites I, II, and III shows decreased diversity. Decreased diversity has been used as an indicator of water quality deterioration. Site IV shows more diversity than the remaining Sites I, II, and III. The considerably much higher diversity at Site IV was probably due to higher nutrient availability and DO contents. Communities with a high diversity use the available energy efficiently and have high stability. In other words, they can resist the adversities of changed environment in a better way. Rai (2011) observed an increasing trend in diversity with dilution from the effluent discharge points. Further, it has been suggested that polluted systems display a reduction in diversity indices. Decrease in diversity indices at polluted regions may be also due to cumulative effects of the various effluents. Staub et al. (1970) proposed a scale of pollution in terms of species diversity, that is, 3.0–4.5, clear; 2.0–3.0, light pollution; 1.0–2.0, moderate pollution; and 0.0–1.0, heavy pollution.

The Sorenson similarity index was calculated between the different sites having common or varying wetland plants/macrophytes. The similarity index may successfully be used for two purposes, that is, whether two or more stands belong to the same community or to determine the extent of resemblance between different communities (Rai, 2011).

PHYTOREMEDIATION OF A RAMSAR SITE (BIODIVERSITY HOTSPOT): CONCLUDING REMARKS

E. crassipes is the top accumulator of Fe in the phytoremediation experiment at all concentrations, that is, 1, 3, and 5 mg for 4, 8, and 12 days. Jayaweera et al. (2008) also reported very high accumulation of Fe in *E. crassipes* biomass under varying nutrient conditions and measured the highest phytoremediation efficiency of 47% during optimum growth at the sixth week with a highest accumulation of 6707 Fe mg/kg dry weight. Furthermore, previous studies showed that hyacinth roots form plaques of $Fe(OH)_3$ by diffusing photosynthetically produced O_2 to the rhizosphere to avoid the formation of H_2S and to counteract Fe^{2+} and Mn^{2+} toxicity at lower DO levels (Vesk and Allaway, 1997; Vesk et al., 1999).

E. crassipes and *P. stratiotes* have been mostly studied for their tendency to bioaccumulate and biomagnify the heavy metal contaminants present in water bodies. These plants were studied to evaluate their efficiency to remove heavy metals from the effluents of a steel foundry located in Hayatabad, Pakistan (Aurangzeb et al., 2014). The potential application of *E. crassipes* in the removal of heavy metals from water was discovered in the early 1980s (Govindaswamy et al., 2011). It accumulates metals, and as the recycling process is run by photosynthetic activity and biomass growth, it is a sustainable and cost-efficient process (Garbisu et al., 2002; Lu et al., 2004; Bertrand and Poirier, 2005). Due to the exotic invasive nature and rapid decomposition of the water hyacinth in comparison with other plants, it has been reported that the growth of this plant poses a problem to the functioning of an aquatic ecosystem, for example, constructed wetlands (Khan et al., 2000; Rai, 2011, 2012). However, it is one of the most suitable plants for phytoremediation used by various researchers and scientists. *E. crassipes* that has colonized natural wetland systems could serve as "nature's kidneys" for proper effluent treatment to preserve the earth's precious water resources from pollution (Malik, 2007). The application of *E. crassipes* as a cleaning agent for phytoremediation is a very useful way of preserving aquatic bodies from many various pollutants, especially heavy metals, because of its easy availability, low cost, effectiveness, and eco-friendliness.

E. crassipes is used in various polluted sites such as wetlands, rivers basins, ponds, ditches, sewages, industrial effluents, and landfills for remediation purposes. *E. crassipes* as a common plant for the removal of Fe, Pb, Cu, Zn, Hg, Cd, Cr, and Mn (Tiwari et al., 2007; Kumar et al., 2008; Rai, 2009; Rai et al., 2010; Chatterjee et al., 2011; Fawzy et al., 2012; Padmapriya and Murugesan, 2012; Sasidharan et al., 2013; Mishra et al., 2013). A study carried out to assess the growth of *E. crassipes* and its ability to accumulate Cu from polluted water with high Cu concentration and mixture of other contaminants under short-term exposure, to use this plant for remediation of highly contaminated sites, found high accumulation and less translocation of Cu in leaves than in roots (Melignani et al., 2015). *E. crassipes* plants were also exposed to varying concentration of Hg as seedlings under hydroponic system to investigate accumulated mercury level, antioxidant defence mechanisms, growth patterns changes, and damage of DNA from the exposure effect (Malar et al., 2015). *E. crassipes* is also an As hyperaccumulator that can take in As, tolerate its presence, and accumulate it in its biomass (Tiwari et al., 2014). *E. crassipes* reduces levels of heavy metals in acid mine water with little sign of toxicity (Falbo and Weaks, 1990).

E. crassipes is very effective in removing nutrients from wetland systems (Wolverton and McDonald, 1979; Trivedy and Pattanshetty, 2002; Jayaweera and Kasturiarachchi, 2004; Cristina et al., 2009). Many researchers have studied *E. crassipes* with wastewater containing high nutrients (Gamage and Yapa, 2001; Jayaweera and Kasturiarachchi, 2004; Tripathi and Upadhyay, 2003; Rai, 2011; Singh and Rai, 2016). The domestic sewage purification potential of this plant was studied using many parameters, including BOD, chemical oxygen demand, fecal coliform count, nitrate and phosphate concentrations, pH value, heavy metal content, turbidity, odor, and color (Alade and Ojoawo, 2009). Moreover, *E. crassipes* has the capacity to remove nutrients and heavy metals from leachate from landfill, minimizing pollution to an acceptable level (Akinbile and Yusoff, 2012; Rai, 2011; Singh and Rai, 2016). Significant results were obtained for the physiological response of *E. crassipes* to the combined exposure of excess nutrients and Hg (Cristina et al., 2009). A study also revealed that *E. crassipes* grown in nutrient-poor conditions is ideal for removing Fe from wastewater with a hydraulic retention time of approximately 6 weeks (Jayaweera et al., 2008). The plant is also effective for purifying wastewater from an intensive duck farm during its growing season, and the harvested plant had also an excellent performance as duck feed (Jianbo et al., 2008). Also, the plant can survive extremely eutrophic water with anaerobically digested flushed dairy manure wastewater (Sooknah and Wilkie, 2004).

P. stratiotes has also been extensively used for phytoremediation (Quian et al., 1999; Skinner et al., 2007). It was also used in lab experiments for the removal of heavy metals (Miretzky et al., 2010). *P. stratiotes* was helpful in the removal of some heavy metals from industrial effluent and was found to be the best phytoremediator for Pb and Cu as it was successful in removing 70.7% and 66.5% of these metals, respectively (Lone et al., 2008). Similar findings were also reported by Lu (2011) while working on lettuce that was a hyperaccumulator of Fe, Pb, and Cu. The same kind of findings were also reported by Mokhtar et al. (2011).

PHYTOCHEMICAL COMPOSITION OF WETLAND PLANTS

E. crassipes, being a fast-growing plant, is used for the rapid removal of various kinds of pollution in water, resulting in positive outcomes. The plant was evaluated for its possible potential for heavy metal accumulation, which resulted in the discovery of high cellulose content and its functional groups including amino ($-NH_2$), carboxyl ($-COO^-$), hydroxyl ($-OH^-$), and sulfhydryl ($-SH$), showing high tolerance and affinity toward heavy metal adsorption (Patel, 2012). *E. crassipes* contains many phytochemicals such as amino acids, including glutamic acid, threonine, leucine, lysine, methionine, tryptophan, tyrosine, and valine, and flavonoids, which include apigenin, azaeleatin, chrysoeriol, gossypetin, kaempferol, luteolin, oientin, and tricin (Nyananyo et al., 2007).

The *P. stratiotes* plant extracts consist of various alkaloids, glycosides, flavonoids, and phytosterols. The leaf and stem extracts consist of 92.9% H_2O, 1.4% proteins, 0.3% fats, 2.6% carbohydrates, 0.9% crude fiber, and 1.9% minerals (mostly potassium and phosphorus). The leaves are rich in vitamins A and C; stigmasterol, stigmasteryl, stigmasterate, and palmitic acids are found in abundance (Khan et al. 2014). 2-di-cgl-cosy-flavones of vicenin and lucenin types, anthocyanin cyanidin-3-glucoside, luteolin-7-glucoside, and mono-*C*-glycosyl flavones vitexin and orientin have also been isolated from the plant (Khare, 2005). Stratioside II (a new C13 norterpene glucoside) is the major component of this plant. The leaves are rich in proteins, essential amino acids, stigmatane, sitosterol acyl glycosides. and minerals (Ghani, 2003). Vicenin, an anticancer agent (Nagaprashantha et al., 2011), and cyanidin-3-glucoside (an anthocyanin), are present (Rastogi and Mehrotra, 1993).

BIOFUEL PRODUCTION

The biogas produced from the organic matter of plants serves as a cheap mode of replaceable biofuel from the present petroleum fuels. Bhattacharya and Kumar (2010) listed the attributes of an ideal biofuel crop. They are as follows:

1. The plant should be naturally grown vegetation, preferably perennials.
2. The plant should have a high cellulose content with low lignin per unit volume of dry matter.
3. The plant should be easily degradable.
4. The plant should not compete with arable crop plants for space, light, and nutrients.
5. The plant should resist pests, insects, and diseases.
6. The plant should not be prone to genetic pollution by crossbreeding with cultivable food crops.

E. crassipes is abundantly available, a perennial, a noncrop, is biodegradable, and has a high cellulose content, which fulfill all the criteria for bioenergy production (Patel, 2012). The cellulose and hemicellulose contents of the plant are more easily converted to fermentable sugar, which results in an enormous amount of utilizable biomass for the biomass industry (Bhattacharya and Kumar, 2010). *E. crassipes* mixed with animal waste yields better biogas (Kumar, 2005) and the obtained sludge feed mixed with nitrogen, phosphorus, and potassium content can be utilized as good manure (Malik, 2007). Microbes take a great role in decomposing the organic matter of plants. Patel et al. (1992) studied the stimulation of microbial activity to increase the biogas production using different biological and chemical additives. Also, the fermentation of the plants' fermentation with the help of methanogens in a reactor turns out to be positive, producing good biofuel (Chanakya et al., 1993).

OTHER UTILITIES

E. crassipes and *P. stratiotes* are multiutility plants that can take different roles from different utility perspectives. They can be used to make traditional baskets and for weaving purposes (Jafari, 2010). Moreover, coasters, place mats, mats, shoes, sandals, bags, wallets, vases, and so on can be crafted from their dried petioles (Patel, 2012). Pulp material extracted from the *E. crassipes* plant has the potential for producing greaseproof paper (Goswami and Saikia, 1994). The plant can also be used for oil sorption in a wide range of temperatures, and the sorbed oils can be recovered (Yang et al., 2014).

Although *E. crassipes* and *P. stratiotes* are potent bio-agents for phytoremediation, their use being suspected by several ecologists in view of their invasive nature which may adversely affect ecosystem functioning of wetlands. However, taking the positive advantages of these plants, technologies have been developed for use in the

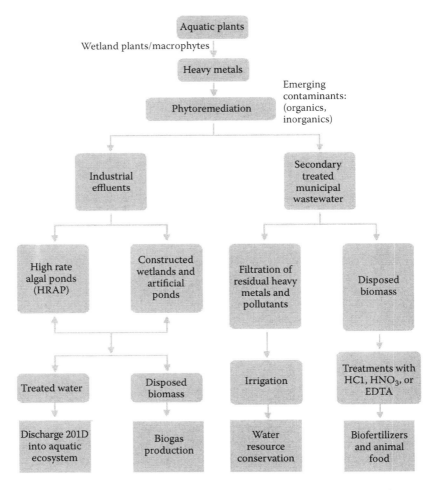

FIGURE 7.6   Eco-sustainable mode for the removal of emerging contaminants (heavy metals) by using wetland plants/macrophytes. (After Rai, 2009; Rai, P.K., *Heavy Metal Pollution and Its Phytoremediation through Wetland Plants*, New York: Nova Science Publisher, 2011; Rai, 2012.)

accumulation and absorption of heavy metals and other nutrients under phytoremediation from aquatic bodies, biofuel and biogas production through fermentation and decomposition, and many more beneficial uses. As the plants have infested every corner and every side of the world, the utilization of these plants should be done from every angle, seeing the positive attributes. This will help in controlling the infestation of these plants and help in making the most of the positive attributes of the *E. crassipes* and *P. stratiotes* in eco-friendly and sustainable ways.

Finally, in Figure 7.6, I illustrate a self-explanatory sustainable way or hypothetical model for removal of emerging contaminants (heavy metals) by using wetland plants/macrophytes.

REFERENCES

Akinbile CO, Yusoff MS. 2012. Water hyacinth (*Eichhornia crassipes*) and lettuce (*Pistia stratiotes*) effectiveness in aquaculture wastewater treatment in Malaysia. *Int J Phytoremediation* 14:201–211.

Aurangzeb N, Nisa S, Bibi Y, Javed F, Hussain F. 2014. Phytoremediation potential of aquatic herbs from steel foundry effluent. *Bra J Chem Eng* 31(4):881–886.

Banerjee US, Gupta S. 2010. Seasonal assessment of irrigation water suitability of river Damodar in West Bengal, India. *J Crops Weeds* 6(1):6–12.

Bertrand M, Poirier I. 2005. Photosynthetic organisms and excess of metals. *Phtosynthetica* 43:345–353.

Bhattacharya A, Kumar P. 2010. Water hyacinth as a potential biofuel crop. *Electron J Environ Agri Food Chem* 9:112–122.

Chanakya HN, Borgaonkar S, Meena G, Jagadish KS. 1993. Solid-phase biogas production with garbage or water hyacinth. *Bioresour Technol* 46:227–231.

Chatterjee S, Chetia M, Singh L, Chattopadyay B, Datta S, Mukhopadhyay SK. 2011. A study on the phytoaccumulation of waste elements in wetland plants of a Ramsar Site in India. *Environ Monit Assess* 178:361–371.

Cristina C, Santiago IT, Jose LA, Jordi B, Anna F. 2009. Physiological responses of *Eichhornia crassipes* [Mart.] Solms to the combined exposure to excess nutrients and Hg. *Bra J Plant Physiol* 21:1–12.

Falbo MB, Weaks TE. 1990. A comparison of *Eichhornia crassipes* (Pontederiaceae) and *Sphagnum quinquefarium* (Sphagnaceae) in treatment of acid mine water. *Econ Bot* 44:40–49.

Fawzy MA, Badr NES, Khatib AE, Kassem AAE. 2012. Heavy metal biomonitoring and phytoremediation potentialities of aquatic macrophytes in River Nile. *Environ Monit Assess* 184:1753–1771.

Gamage NS, Yapa PAJ. 2001. Use of water hyacinth [*Eichhornia crassipes* (Mart) Solms] in treatment systems for textile mill effluents—A case study. *J Natl Sci Found Sri Lanka* 29:15–28.

Garbisu C, Hernandez-Allica J, Barrutia O, Alkortaand I, Becerril JM. 2002. Phytoremediation: A technology using green plants to remove contaminants from polluted areas. *Rev Env Health* 17(3):173–188.

Ghani A. 2003. *Medicinal Plants of Bangladesh with Chemical Constituents and Uses*. 2nd ed. Dhaka: Asiatic Society of Bangladesh.

Goswami T, Saikia CN. 1994. Water hyacinth—A potential source of raw material for grease-proof paper. *Bioresour Technol* 50:235–238.

Govindaswamy S, Schupp DA, Rock SA. 2011. Batch and continuous removal of arsenic using hyacinth roots. *Int J Phytoremediation* 13:513–527.

Gurzau ES, Neagu C, Gurzau AE. 2003. Essential metals—Case study on iron. *Ecotox Enviro. Saf* 56(1):190–200.

Hacioglu N, Dulger B. 2009. Monthly variation of some physico-chemical and microbiological parameters in Biga Stream (Biga, Canakkale, Turkey). *AfrJ Biotechnol* 8(9):1929–1937.

Jayaweera MW, Kasturiarachchi JC. 2004. Removal of nitrogen and phosphorus from industrial wastewaters by phytoremediation using water hyacinth (*Eichhornia crassipes* [Mart.] Solms). *Water Sci Technol* 50:217–225.

Jafari N. 2010. Ecological and socio-economic utilization of water hyacinth (Eichhornia crassipes Mart Solms). *J. Appl. Sci. Environ. Manage* 14(2):43–49.

Jayaweera MW, Kasturiarachchi JC, Kularatne RK, Wijeyekoon SL. 2008. Contribution of water hyacinth (*Eichhornia crassipes* (Mart.) Solms) grown under different nutrient conditions to Fe-removal mechanisms in constructed wetlands. *J Environ Manage* 87(3):450–460.

Jianbo L, Zhihui F, Zhaozheng Y. 2008. Performance of a water hyacinth (*Eichhornia crassipes*) system in the treatment of wastewater from a duck farm and the effects of using water hyacinth as duck feed. *J Environ Sci* 20:513–519.

Kakati GN, Bhattacharya KG. 1990. Trace metal in surface water of Greater Gauhati. *Indian J Environ Health* 32:197.

Kangabam RD et al. 2017. Development of a water quality index (WQI) for the Loktak Lake in India. *Appl Water Sci* 7:2907–2918.

Kataria HC, Iqbal SA. 1995. Chloride content in borewells (ground water) of Bhopal City. *Asian J Chem Rev* 6:35–38.

Khan AG, Kuek C, Chaudry TM, Khoo CS, Hayes WJ. 2000. Role of plants, mycorrhizae and phytochelators in heavy metals contaminated land remediation. *Chemosphere* 41:197–207.

Khan MA, Marwat KB, Gul B, Wahid F, Khan H, Hashim S. 2014. *Pistia stratiotes* L. (Araceae): Phytochemistry, use in medicines, phytoremediation, biogas and management options. *Pak J Bot* 46(3):851–860.

Khan MAG and Choudhary SH. 1994. Physical and chemical limnology of lake Kaptai: Bangladesh. *Tropical Ecology* 35(1): 35–51.

Khan RM, Jadhav MJ, Ustad IR. 2012. Physicochemical analysis of Triveni Lake water of Amravati District in (MS) India. *Biosci Discov* 3(1):64–66.

Khare CP. 2005. *Encyclopedia of Indian Medicinal Plants*. Berlin: Springer-Verlag.

Khare KC, Jadhav MS. 2008. Water quality assessment of Katraj Lake, Pune (Maharastra, India): A case study. In *The 12th World Lake Conference*, ed. M Sengupta, R Dalwani R, 292–299.

Kosygin L, Dhamendra H. 2009. Ecology and conservation of Loktak Lake, Manipur: An overview. In *Wetlands of North East India Ecology, Aquatic Bioresources and Conservation*, ed. K Laishram, 1–20. New Delhi: Akansha Publishing House.

Kulkarni JR, Shrivastava VS. 2000. Physical and chemical investigative for the assessment of pollution in and around industrial area. *Indian J Env Prot* 20(4):252–256.

Kumar A. 2000. A quantitative study of the pollution and physico-chemical conditions of the river Mayurakshi in Santhal Pargana, Bihar. In *Pollution and Biomonitoring of Indian Rivers*, ed. RK Trivedy, 246–251. Jaipur: ABD.

Kumar JIN, Soni R, Kumar RN, Bhatt I. 2008. Macrophytes in phytoremediation of heavy metal contaminated water and sediments in Periyej Community Reserve, Gujarat, India. *Turk J Fish Aquat Sci* 8:193–200.

Kumar S. 2005. Studies on efficiencies of bio-gas production in anaerobic digesters using water hyacinth and night-soil alone as well as in combination. *Asian J Chem* 17:934–938.

Lone MI et al. 2008. Phytoremediation of heavy metal polluted soils and water: Progresses and perspectives. *J Zhejiang Univ Sci B* 9(3):210–220.

Lu Q et al. 2011. Uptake and distribution of metals by water lettuce (Pistia stratiotes L.). *Environ Sci Pollut Res* 18:978–986.

Lu X, Kruatrachue M, Pokethitiyook P, Homyok K. 2004. Removal of cadmium and zinc by water hyacinth, *Eichhornia crassipes. Sci Asia* 30:93–103.

Maine MA, Duarte MV, Sue NL. 2001. Cadmium uptake by floating macrophytes. *Water Res* 35:2629–2634.

Malar S, Sahi SV, Favas PJC, Venkatachalam P. 2015. Mercury heavy-metal-induced physio-chemical changes and genotoxic alterations in water hyacinths [*Eichhornia crassipes* (Mart.)]. *Environ Sci Pollut Res* 22:4597–4608.

Malik A. 2007. Environmental challenge *vis a vis* opportunity: The case of water hyacinth. *Environ Int* 33:122–138.

Manivasakam N. 1986. *Physico-chemical Examination of Water, Sewage and Industrial Effluents*, 2nd revised and enlarged ed. Meerut: Pragati Prakashan.

Melignani E, de Cabo LI, Faggi AM. 2015. Copper uptake by *Eichhornia crassipes* exposed at high level concentrations. *Environ Sci Pollut Res* 22:8307–8315.

Mishra S, Mohanty M, Pradhan C, Patra HK, Das R, Sahoo S. 2013. Physico-chemical assessment of paper mill effluent and its heavy metal remediation using aquatic macrophytes— A case study at JK Paper mill, Rayagada, India. *Environ Monit Assess* 185:4347–4359.

Mittal SK, Verma N. 1997. Critical Analysis of Ground Water Quality Parameters. *Indian J Env Prot* 17(6):426–429.

Miretzky P, Saralegui A, Cirelli AF. 2010. Quantitative assessment of worldwide contamination of air, water and soil by trace elements. *Nature* 279:409–411.

Mokhtar H, Norhashimah M, Fera F. 2011. Phytoaccumulation of copper from aqueous solutions using *Eichhornia crassipes* and *Centella asiatica. Int J Environ Sci Develop* 2:205–210.

Nagaprashantha LD, Vatsyayan R, Singhal J, Fast S, Roby R. 2011. Anti-cancer effects of novel flavonoid vicenin-2 as a single agent and in synergistic combination with docetaxel in prostate cancer. *Biochem Pharmacol* 82:1100–1109.

Naik S. 2005. Studies on pollution status of Bondamunda area of Rourkela industrial complex. PhD dissertation, National Institute of Technology, Rourkela, India.

Nybakken JW. 1997. *Marine Biology: An Ecological Approach*, 4th ed. Reading, MA: Addison-Wesley Educational Publishers Inc.

Nyananyo BL, Gijo A, Ogamba EN. 2007. The physicochemistry and distribution of water hyacinth (*Eichhornia crassipes*) on the river Nun in the Niger Delta. *J Appl Sci Environ Manage* 11:133–137.

Odjegba VJ, Fasidi IO. 2004. Accumulation of trace elements by *Pistia stratiotes*: Implications for phytoremediation. *Ecotoxicology* 13:637–646.

Padmapriya G, Murugesan AG. 2012. Phytoremediation of various heavy metals (Cu, Pb and Hg) from aqueous solution using water hyacinth and its toxicity on plants. *Int J Environ Biol* 2(3):97–103.

Palharya JP, Siriah VK, Malariya S. 1993. *Environmental Impact of Sewage and Effluent Disposal on the River System*. New Delhi: Ashish.

Patel S. 2012. Threats, management and envisaged utilizations of aquatic weed *Eichhornia crassipes*: An overview. *Rev Environ Sci Biotechnol* 11:249–259.

Patel V, Patel A, Madamwar D. 1992. Effects of adsorbents on aerobic digestion of water hyacinth-cattle dung'. *Bioresour Technol* 40:179–181.

Quian JH, Zayed A, Zhu YL et al. 1999. Phytoaccumulation of tace elements by wetland plants, III: Uptake and accumulation of ten trace elements by twelve plant species. *J Environ Qual* 28:1448–1455

Rai PK. 2008. Heavy-metal pollution in aquatic ecosystems and its phytoremediation using wetland plants: An ecosustainable approach. *Int. J. Phytoremediation.* 10(2): 133–160.

Rai PK. 2009. Heavy metal phytoremediation from aquatic ecosystems with special reference to macrophytes. *Cri Rev Env Sci Tec* 39(9):697–753.

Rai PK. 2010a. Microcosm investigation on phytoremediation of Cr using Azolla pinnata. *Int J Phytoremediation* 12:96–104.

Rai PK. 2010b. Phytoremediation of heavy metals in a tropical impoundment of industrial region. *Environ Monit Assess* 165:529–537.

Rai PK. 2010c. Seasonal monitoring of heavy metals and physico-chemical characteristics in a lentic ecosystem of sub-tropical industrial region, India. *Environ Monit Assess* 165:407–433.

Rai PK. 2010d. Heavy metal pollution in lentic ecosystem of sub-tropical industrial region and its phytoremediation. *Int J Phytoremediation* 12(3):226–242.

Rai PK. 2011. *Heavy Metal Pollution and Its Phytoremediation through Wetland Plants.* New York: Nova Science Publisher.

Rai PK. 2012. An Eco-sustainable green approach for heavy metals management: Two case studies of developing industrial region. *Environ Monit Assess* 184:421–448.

Rastogi RP, Mehrotra BN. 1993. *Compendium of Indian Medicinal Plants.* New Delhi: Lucknow and Publications & Information Directorate.

Sasidharan NK, Azim T, Devi DA, Mathew S. 2013. Water hyacinth for heavy metal scavenging and utilization as organic manure. *Indian J Weed Sci* 45(3):204–209.

Singh KK, Sharma BM, Usha Kh. 2010. Ecology of Kharungpat lake, Thoubal, Manipur, India: Part-I Water quality status. *The Ecoscan* 4(2–3):241–245.

Singh MM, Rai PK. 2016. Microcosm investigation of Fe (iron) removal using macrophytes of Ramsar Lake: A phytoremediation approach. *Int J Phytoremediation* 18(12):1231–1236

Skinner K, Wright N, Goff EP. 2007. Mercury uptake and accumulation by four species of aquatic plants. *Environ Pollut* 145:234–237.

Sooknah RD, Wilkie AC. 2004. Nutrient removal by floating aquatic macrophytes cultured in anaerobically digested flushed dairy manure wastewater. *Ecol Eng* 22:27–42.

Staub R, Appling JW, Hofsteiler AM, Hess IJ. 1970. The effect of industrial waste of Memphis and Shelby country on primary plankton producers. *Bioscience* 20:905–912.

Tiwari S, Dixit S, Verma N. 2007. An effective means of biofiltration of heavy metal contaminated water bodies using aquatic weed *Echhornia crassipes*. *Environ. Monit Assess* 129:253–256.

Tiwari S, Sarangi BK, Pandey RA. 2014. Efficacy of three different plants species for arsenic phytoextraction from hydroponic system. *Environ En. Res* 19:145–149.

Tripathi BD, Upadhyay AR. 2003. Dairy effluent polishing by aquatic macrophytes. *Water Air Soil Pollut* 9:377–385.

Trivedy RK, Goel PK. 1986. *Chemical and Biological Methods for Water Pollution Studies.* Karad, India: Environmental Publications.

Trivedy RK, Pattanshetty SM. 2002. Treatment of dairy waste by using water hyacinth. *Water Sci Technol* 45:329–334.

Umavathi S, Logankumar K. 2010. Physicochemical and nutrient analysis of Singanallur Pond, Tamil Nadu (India). *Pollut Res* 29(2):223–229.

Vesk PA, Allaway WG. 1997. Spatial variation of copper and lead concentrations of water hyacinth plants in a wetland receiving urban run-off. *Aquat Bot* 59:33–44.

Vesk PA, Nockold CE, Aaway WG. 1999. Metal localization in water hyacinth roots from an urban wetland. *Plant Cell Environ* 22:149–158.

Wolverton BC, McDonald RC. 1979. Water hyacinth: From prolific pest to potential provider. *Ambio* 8:2–9.

Yang X, Chen S, Zhang R. 2014. Utilization of two invasive free-floating aquatic plants (*Pistia stratiotes* and *Eichhornia crassipes*) as sorbents for oil removal. *Environ Sci Pollut Res* 21:781–786.

Zafar A, Sultana N. 2008. Seasonal analysis in the water quality of the river Ganga. *J Curr Sci* 12(1):217–220.

8 Wastewater Treatment with Green Chemical Ferrate: An Eco-Sustainable Option

INTRODUCTION

Over the past few decades, diverse chemicals and materials such as mono- and bimetallic nanoparticles, metal oxides, and zeolites have been used for soil and groundwater remediation. In this context, the advent of green chemistry principles and approach led to a paradigm shift in the use of chemicals in the remediation of emerging contaminants present in surface water/groundwater. In the recent past, there was a general assumption that chemicals would pose several side effects to the total environment; however, certain green and eco-friendly chemicals have revolutionized environmental remediation perspectives with remarkable success. To this end, green chemistry is the design of chemical products and processes that reduce or eliminate the use and generation of hazardous substances unlike conventional approaches (Warner et al., 2004; Kirchhoff, 2005; Rai, 2018). The Environmental Protection Agency defines green chemistry as the use of chemistry for the prevention of pollution at the molecular level (García-Serna et al., 2007; Rai, 2012). There are three prime focusing areas of green chemistry: (1) the use of alternative synthetic pathways, (2) the use of alternative reaction conditions, and (3) the design of safer chemicals that are less toxic than current alternatives or inherently safer with regard to accident potential (García-Serna et al., 2007).

Over the past decades, we have witnessed the occurrences of diverse emerging contaminants (such as heavy metals, volatile organic carbons, pesticides, personal care products [PCPs], pharmaceuticals, organics, and pathogenic microbes) that perturbed all different compartments of the environment. As such, the degradation of groundwater and surface water by diverse emerging contaminants is recognized as a global problem that is still in quest for affordable (cost-effective) and eco-technological solution. To this end, the prospect of green chemicals such as ferrate was tested to attain eco-management of emerging contaminants in surface water/groundwater (Rai, 2012, 2017). In line with this, the Green Lead Project of Denmark was a good example that offered the provision of mining, processing, transporting, treating, manufacturing, storing, using, and recycling of lead (Pb) with zero harm (from its exposure) to both people and the environment (Roche and Toyne, 2004). Similarly, catalysis has been one of the key technologies employed to achieve such objectives in the practice of sustainable (green) chemistry (Centi and Perathoner, 2003).

TABLE 8.1
Comparison of Reduction Potential of Fe(VI) with Those of Oxidants Used in Water Treatment

S. No.	Oxidant	pH	Reaction	Potential (V)
1	Ferrate	Acidic	$FeO_4^{2-} + 8H^+ + 3e^- \leftrightarrow Fe^{3+} + 4H_2O$	2.20
		Basic	$FeO_4^{2-} + 4H_2O + 3e^- \leftrightarrow Fe(OH)_3 + 5OH^-$	0.70
2	Hypochlorite	Acidic	$HClO + H^+ + 2e^- \leftrightarrow 2Cl^- + H_2O$	1.48
		Basic	$ClO^- + H_2O + 2e^- \leftrightarrow 2Cl^- + 2OH^-$	0.84
3	Ozone	Acidic	$O_3 + 2H^+ + 2e^- \leftrightarrow O_2 + H_2O$	2.08
		Basic	$O_3 + H_2O + 2e^- \leftrightarrow O_2 + 2OH^-$	1.24
4	Hydrogen peroxide	Acidic	$H_2O_2 + 2H^+ + 2e^- \leftrightarrow 2H_2O$	1.78
		Basic	$H_2O_2 + 2e^- \leftrightarrow 2OH^-$	0.88
5	Permanganate	Acidic	$MnO_4^- + 4H^+ + 3e^- \leftrightarrow MnO_2 + 2H_2O$	1.68
		Basic	$MnO_4 + 2H_2O + 3e^- \leftrightarrow MnO_2 + 4OH^-$	0.59

The ferrate(VI) ion, with the molecular formula, FeO_4^{2-}, is a very strong oxidant. Under acidic conditions, the redox potential of ferrate(VI) ions (2.2 V) is the strongest of all the oxidants/disinfectants practically used for water and wastewater treatment (Table 8.1); note that it is greater than that of ozone (2.0 V). Moreover, during the oxidation/disinfection process, ferrate(VI) ions will be reduced to Fe(III) ions or ferric hydroxide. As such, this simultaneously generates a coagulant in a single dosing and mixing unit process (Jiang and Lloyd, 2002). The FeO_4^{2-} ion has a tetrahedral structure similar to its geometry in the solid state; these four Fe–O bonds are equivalent with covalent characteristics (Jiang and Lloyd, 2002; Hoppe et al., 1982). The thermodynamic constants of potassium ferrate were first measured/calculated by Wood (1958). There are two major methods for characterizing ferrate(VI) salts in practice, such as the volumetric titration method and spectroscopy method (Jiang and Lloyd, 2002).

Ferrate(VI) showed promising behavior toward the oxidation of several inorganic and organic emerging contaminants (Yngard et al., 2007). The growing body of research articles revealed that due to its unique properties (viz., strong oxidizing potential and simultaneous generation of ferric coagulating species), ferrate(VI) salt can disinfect microorganisms, degrade and/or oxidize organic and inorganic impurities (at least partially), and remove suspended/colloidal particulate materials in a single dosing and mixing unit process (Jiang and Lloyd, 2002). In view of these unique properties of ferrate, this chapter aims to provide a comprehensive review on the preparation, properties, and role of ferrate in helping resolve multifaceted environmental challenges.

PREPARATION OF FERRATE

The synthesis of ferrate(VI) has been investigated systematically over the past century. However, such efforts date back to Stahl in 1702, when he first conducted an experiment detonating a mixture of saltpeter and iron filings by dissolving the molten residue in water.

Jiang and Lloyd (2002) in their extensive review reported the methodology involved in ferrate preparation. There are three methods for the preparation of ferrate(VI), namely, (1) dry oxidation by heating/melting various iron oxide-containing minerals under conditions of strong alkaline and oxygen flow, (2) electrochemical method by anodic oxidation using iron or alloy as anode and NaOH or KOH as electrolyte, and (3) wet oxidation by oxidizing a Fe(III) salt at a strong alkaline condition and using hypochlorite or chlorine as the oxidant (see Jiang and Lloyd [2002] and Rai [2017] for details).

GENERAL ASPECTS OF WASTEWATER TREATMENT BY FERRATE

Some approaches used to combat microbial agents (e.g., use of chlorine) are found as incapable of destroying anthrax and certain microbes while leaving disinfectant by-products, which may have serious human health implications (Sharma, 2004). Alternative disinfectants (e.g., bromine, iodine, chlorine dioxide, and ozone) have thus been considered to replace chlorine. However, those replacements are seen to form a range of other by-products, which are also toxic to some extent to the human population and to aquatic life (Jiang, 2007). Fe(V) in general and Fe(VI) in particular may provide a better solution to these issues as quite reliable water and wastewater treatment tools (Sharma, 2004).

Coagulation and oxidation/disinfection are two important unit processes for wastewater treatment loaded with diverse emerging contaminants (Jiang, 2006). For coagulation and disinfection of sewage, it has been demonstrated that potassium ferrate(VI) is more efficient in removing organic contaminants, chemical oxygen demand (COD), and bacteria in comparison with the other two conventional coagulants, specifically ferric sulfate and aluminum sulfate, when applied at the same dosage. Moreover, it has been observed that potassium ferrate(VI) produced less sludge volume while removing more contaminants to make subsequent sludge treatment easier.

Coagulation destabilizes colloidal contaminants and transfers small particles into large aggregates and adsorbs dissolved organic materials onto the aggregates, which can then be removed by sedimentation and filtration. Disinfection is designed to kill harmful organisms (e.g., bacteria and viruses), and oxidation is used to degrade various organic contaminants. The ferrate(VI) ion is also a coagulant; during the oxidation/disinfection process, ferrate(VI) ions are reduced to Fe(III) ions or ferric hydroxide, and this simultaneously generates a coagulant in a single dosing and mixing unit process (Jiang and Lloyd, 2002; Jiang, 2006). More importantly, ferrate(VI) is also an environmentally friendly treatment chemical that will not produce any harmful by-products in the treatment process.

In the context of the remediation of emerging contaminants, a vast number of research observations have shown that potassium ferrate(VI) can perform superiorly in degrading various synthetic and natural organic contaminants (Jiang, 2006; Norcross et al., 1997; Bartzatt and Carr., 1986; Sharma and Bielski, 1991; Bielski et al., 1994; Gulyas, 1997; White and Franklin, 1998; Sharma, 2002; Jiang and Wang, 2003a; Rai, 2012, 2017), inactivating harmful microorganisms (Jiang et al., 2006; Murmann and Robinson, 1974; Gilbert et al., 1976; Schink and Waite, 1980; Kazama, 1995; Jiang and Wang, 2003b; Rai, 2017), and coagulating colloidal particles and

TABLE 8.2
Fe(VI) Oxidation of Emerging Contaminants in the Total Environment at 25°C

S. No.	Emerging Contaminant in the Environment	pH	K (M/s)	References
1	Hydrogen sulfide	9	7.4×10^5	Sharma et al. (1997), Sharma (2002)
2	p-Hydroquinone	9	2.0×10^5	Bielski (1991), Sharma (2002)
3	2-Mercaptoethanesulfonic acid	9	3.0×10^4	Read et al. (1998b), Sharma (2002)
4	2-Mercaptobenzoic acid	10	2.5×10^4	Read et al. (1998b), Sharma (2002)
5	3-Mercaptopropionic acid	9	1.3×10^4	Read and Wyand (1998), Sharma (2002)
6	Methyl hydrazine	9	9.8×10^3	Johnson and Hornstein (1994)
7	Aniline	9	6.2×10^3	Sharma and Hollyfield (1995)
8	Hydrazine	9	5.6×10^3	Johnson and Hornstein (1994)
9	Thioacetamide	9	5.5×10^3	Sharma et al. (2000), Sharma (2002)
10	Ferrocyanide	9	4.7×10^3	Johnson and Sharma (1999), Sharma (2002)
11	Thiourea	9	3.4×10^3	Sharma et al. (1999), Sharma (2002)
12	p-Toluidine	9	1.3×10^3	Sharma and Hollyfield (1995)
13	Cysteine	12.4	7.6×10^2	Sharma and Bielski (1991), Sharma (2002)
14	Thiosulfate	9	7.2×10^2	Johnson and Read (1996), Sharma (2002)
15	Glyoxylic acid	8	7.0×10^2	Carr et al. (1995), Sharma (2002)
16	Cyanide	9	6.2×10^2	Sharma et al. (1998), Sharma (2002)
17	Glyoxal	8	3.0×10^2	Carr et al. (1995), Sharma (2002)
18	Dimethylamine	8	2.0×10^2	Carr et al. (1981), Sharma (2002)
19	Benzenesulfinate	9	1.4×10^2	Johnson and Read (1996), Sharma (2002)
20	Methionine	9	1.3×10^2	Sharma and Bielski (1991), Sharma (2002)
21	Cystine	12.4	1.2×10^2	Sharma and Bielski (1991), Sharma (2002)
22	Sarcosine	8	1.2×10^2	Carr et al. (1995), Sharma (2002)
23	Iminodiacetic acid	8	1.0×10^2	Carr et al. (1981), Sharma (2002)
24	Glycine	8	1.0×10^2	Sharma and Bielski (1991), Sharma (2002)
25	Diethylsulfide	8	1.0×10^2	Carr et al. (1995), Sharma (2002)
26	Thiodiethanol	8	1.0×10^2	
27	Phenol	9	8.0×10^1	Carr et al. (1995), Sharma (2002
28	Thioxane	9	5.8×10^1	Read et al. (1998a), Sharma (2002)
29	p-Aminobenzoic acid	9	4.3×10^1	Sharma and Hollyfield (1995)
30	Methylamine	8	4.0×10^1	Carr et al. (1995), Sharma (2002)
31	p-Nitroaniline	9	3.0×10^1	Sharma and Hollyfield (1995)
32	Chloral	8	6.0×10^0	Carr et al. (1995), Sharma (2002)
33	Glycolaldehyde	8	3.0×10^0	Carr et al. (1995), Sharma (2002)
34	Dimethylglycine	8	2.5×10^0	Carr et al. (1995), Sharma (2002)
35	Trimethylaldehyde	8	2.0×10^0	Carr et al. (1995), Sharma (2002)
36	Nitriloacetic acid	8	2.0×10^0	Carr et al. (1995), Sharma (2002)
37	N-methyliminodiacetic acid	8	2.0×10^0	Carr et al. (1995), Sharma (2002)
38	Dimethylsulfoxide	8	1.0×10^0	Carr et al. (1995), Sharma (2002)
39	Nitrite	9	6.2×10^{-1}	Sharma et al. (1998), Sharma (2002)

(Continued)

TABLE 8.2 (CONTINUED)
Fe(VI) Oxidation of Emerging Contaminants in the Total Environment at 25°C

S. No.	Emerging Contaminant in the Environment	pH	K (M/s)	References
40	Diethylamine	8	7.0×10^{-1}	Carr et al. (1995), Sharma (2002)
41	Formaldehyde	8	5.0×10^{-1}	Carr et al. (1995), Sharma (2002)
42	Acetaldehyde	8	4.0×10^{-1}	Carr et al. (1995), Sharma (2002)
43	Formic acid	8	4.0×10^{-1}	Carr et al. (1995), Sharma (2002)
44	Glycolic acid	8	4.0×10^{-1}	Carr et al. (1995), Sharma (2002)
45	Ammonia	9	1.7×10^{-1}	Carr et al. (1995), Sharma (2002)
46	Oxalic acid	8	1.0×10^{-1}	Carr et al. (1995), Sharma (2002)
47	Neopentyl alcohol	8	1.0×10^{-1}	Carr et al. (1995), Sharma (2002)
48	Ethyl alcohol	8	8.0×10^{-2}	Carr et al. (1995), Sharma (2002)
49	Isopropyl alcohol	8	6.0×10^{-2}	Carr et al. (1995), Sharma (2002
50	Ethylene glycol	8	4.0×10^{-2}	Carr et al. (1995), Sharma (2002)
51	Methyl alcohol	8	3.0×10^{-2}	

heavy metals (Jiang et al., 2006; Bartzatt, 1992; Jiang and Lloyd, 2002; Jiang, 2003; Rai, 2017). Table 8.2 cites some example of Fe(VI) oxidation of contaminants at 25°C.

EFFECT OF FERRATE TREATMENT IN TERMS OF COMMON INDICES OF WATER QUALITY PARAMETERS AND EMERGING ENVIRONMENTAL CONTAMINANTS

In recent years, micropollutants have been identified as emerging water contaminants that are still in quest for an eco-technological solution using the green chemistry approach. To this end, a novel technology for enhanced municipal wastewater treatment was assessed based on the dual functions of Fe(VI) in oxidizing emerging water contaminants and removing phosphate by the formation of ferric phosphates. Moreover, Fe(VI) has been known to react with electron-rich organic moieties (ERMs) of emerging water contaminants such as phenols (Rush et al., 1996; Lee et al., 2005), anilines (Hornstein, 1999; Huang et al., 2001; Sharma et al., 2006), amines (Sharma et al., 2006; Lee et al., 2008), and olefins (Hu et al., 2009). Therefore, ERM-containing compounds can be potentially transformed during Fe(VI) oxidation, an inherent chemical attribute of this green multipurpose chemical. This has been demonstrated in a few kinetics studies made previously for the Fe(VI) reaction with emerging contaminants such as steroid estrogens (Lee et al., 2005), sulfonamide antimicrobials (Sharma et al., 2006), and carbamazepine (Hu et al., 2009). To this end, Lee et al. (2009) assessed the potential of Fe(VI) to oxidize selected emerging contaminants and to remove phosphate during enhanced treatment of sewage wastewater in a single treatment step. The second-order rate constant (k) for the reaction of Fe(VI) with selected emerging contaminants and organic model compounds was determined as a function of solution pH (Lee et al., 2009).

The superior performance of potassium ferrate(VI) as an oxidant/disinfectant for environmental remediation has been demonstrated in various research

(Jiang, 2007). Nonetheless, some challenges have existed regarding the implementation of ferrate(VI) technology in full-scale treatment of wastewater and sewage sludge owing to either the instability property of ferrate(VI) solution or the high preparation cost of solid ferrate(VI) (Jiang, 2007). The treatment efficiencies of Fe(VI) and $FeCl_3$ were compared by Stanford et al. (2010). Removal of phosphorus reached 40% with a Fe(VI) dose as low as 0.01 mg/L compared with 25% removal with 10 mg/L of Fe(III). For lower doses (<1 mg/L as Fe), Fe(VI) achieved removal of suspended solids (SS) and COD between 60% and 80%. In contrast, Fe(III) did not perform well in the control sample where no iron chemical was dosed. The ferrate solution was found to be stable for a maximum of 50 min, beyond which Fe(VI) was reduced to less oxidant species. This provided the maximum allowable storage time of the electrochemically produced ferrate(VI) solution. Results demonstrated that low addition of ferrate(VI) should lead to good removal of P, biological oxygen demand (BOD), COD, and SS from wastewater compared with ferric addition (Stanford et al., 2010; Jiang, 2013).

EFFECT OF FERRATE TREATMENT ON EMERGING CONTAMINANTS/MICROPOLLUTANTS: EDCs, PPCPs, SURFACTANTS, AND ORGANIC POLLUTANTS

In the recent context of the rapid pace of urbanization and industrialization, perturbations of different environmental compartments such as soil and groundwater have become a matter of serious concern. To this end, pharmaceutical and personal care products (PPCPs) and endocrine-disrupting chemicals (EDCs) are classified as emerging micropollutants because they may have significant adverse environmental effects, although they are at very low concentration ranges (Horvath and Huszank, 2003; Jiang et al., 2013; Rai, 2017). In the human health sector, pharmaceuticals such as antibiotics, anti-inflammatory drugs, β-blockers, and X-ray contrast media are widely used, which possess the potential to contaminate and pollute environmental compartments such as the soil and groundwater. Hence, these pharmaceuticals and their metabolites are inevitably released into the water bodies through excretion (Comeau et al., 2008) and/or through the discharge of industry effluents and hospital wastewaters (Ternes, 1998).

EDCs are defined as natural and/or synthetic compounds that could affect the endocrine systems of fish and other aqueous animals. In the past two decades, a variety of adverse effects of EDCs on the endocrine systems of animals have been observed (Jiang, 2013; Piva and Martini, 1998; Thorpe et al., 2001). These effects may be cumulative and hence possibly will appear only in subsequent generations, and then the resultant effects may be irreversible, threatening human sustainable development. Most EDCs are synthetic organic chemicals that are introduced to the environment by anthropogenic inputs (e.g., bisphenol A [BPA]). However, they can also be naturally generated estrogenic hormones, for example, estrone (E1) and 17β-estradiol (E2). As such, they are present ubiquitously in all kinds of aquatic environments receiving wastewater effluents (Jiang, 2004). Tables 8.3 and 8.4 show the rate constants and interactions of ferrate with EDCs and PPCPs. Whilst potassium ferrate alone had a minor effect on color, the combination of ferrate with the organic

TABLE 8.3

Ferrate(VI) Removal of Emerging Contaminants Such as EDCs and PPCPs

Group	Contaminant	pH	Temperature (°C)	References
A. Endocrine disruptors	17α-Ethinylestradiol	7		Lee et al. (2005)
	BPA	7	24	Jiang (2007)
	α-Estradiol	7		Lee et al. (2009)
	Phenol	7		
	17β-estradiol	7		
	4-Methylphenol	7		
	Buten-3-ol	7		
B. PPCPs	Atenolol	8		Lee and von Gunten (2010)
	Bisulfite	7	25	Yang et al. (2012)
	Carbamazepine	8		Sharma et al. (2011)
	Ciprofloxacin	7		Noorhasan et al. (2010)
	Enrofloxacine	7		Sharma et al. (2011)
	Ethionine	8	25	Sharma et al. (2006)
	Glycylglycine	7	25	
	Ibuprofen	8		
	Iodide	7	25	
	Sulfamethiozole	7		
	Sulfamethoxazole	7	25	
	Sulfisoxazole	7		

TABLE 8.4

Ferrate(VI) Treatment Performance and Second-Order Rate Constants K (M/s) for Degradation of Selected Emerging Contaminants EDCs and PPCPs

S. No.	Pollutant	Treatment Performance	Reaction Rate Constant of Kinetics K (M/s)	References
1	17α-Ethinyl estradiol, 17β-estradiol, and BPA	pH 8, 25°C, 30 min, [Fe(VI)] > 1 mg/L as Fe, >99% of chemicals removed	pH 7, 25°C: $(6.4–7.7) \times 10^2$	Lee et al. (2005), Jiang et al. (2005)
2	Ibuprofen	Up to 40% reduction	pH 8.0, 25°C: 0.09	Sharma and Mishra (2006)
3	Octylphenols	N.A.	pH 7.0, 25°C, $(1.18 \pm 0.05) \times 10^3$	Anquandah and Sharma (2009)
4	Carbamazepine	N.A.	pH 7.0, 25°C: 70 ± 3	Hu et al. (2009)

polymer allowed good decolorization: this suggested the eventual application of this combined process for reuse of dyeing wastewater, resulting in environmental and economic benefits (Ciabatti et al., 2010).

EDCs and PPCPs have been found incompletely removed in various conventional wastewater treatment plants (WWTPs) (Ying et al., 2008, 2009; Liu et al., 2009; Khetan and Collins, 2007). The presence of high amounts of EDCs and PPCPs in WWTP effluents and in receiving aquatic environments may also pose potential risks to aquatic organisms and human health (Ying et al., 2009; Khetan and Collins, 2007; Witorsch and Thomas, 2010). The treatability of these pollutants with Fe(VI) has been investigated intensively, as well as that of diverse organic compounds: alcohol (Norcross et al., 1997), aliphatic sulfur (Jiang et al., 2007; Bartzatt and Carr, 1986), amino acids (Jiang et al., 2007; Sharma and Bielski, 1991), carboxylic compounds (Jiang et al., 2007; Bielski et al., 1994), organic nitrogen compounds, phenol and its related compounds, recalcitrant organics (Gulyas, 1997), and thiourea (Jiang et al., 2007).

In this context, Yang et al. (2012) investigated the removal efficiencies of 68 selected EDCs and PPCPs spiked in a wastewater matrix by ferrate (Fe(VI)) and further evaluated the degradation of these micropollutants present in secondary effluents of two WWTPs by applying Fe(VI) treatment technology. Fe(VI) treatment resulted in selective oxidation of ERMs of these target compounds such as phenol, olefin, amine, and aniline moieties. However, Fe(VI) failed to react with triclocarban, three androgens, seven acidic pharmaceuticals, two neutral pharmaceuticals, and erythromycin–H_2O.

Further, in the context of other emerging contaminants, that is, surface-active agents, commonly called surfactants, contain a group of organic compounds that have both hydrophobic (e.g., alkyl chain) and hydrophilic groups (e.g., ammonium ion). Surfactants impact on all aspects of our daily life either directly in household detergents and PCPs or indirectly in the production and processing of materials that surround us (Eng et al., 2006). To this end, the major concern arising regarding sewage treatment plants is due to some of the surfactants, especially cationic surfactants that are partially biodegraded and, thus, contaminate the aquatic environment (Eng et al., 2006; Garcia et al., 2000). Therefore, it is possible to speculate that the environmental pollution caused by surfactants is increasing due to daily life applications (Eng et al., 2006, Horvath and Huszank, 2003).

Cyanide, a highly toxic substance and emerging contaminant of environmental concern, is present in wastes from gold refining, metal plating, chemical manufacturing, and iron and steel industries and may be oxidized by Fe(VI) (Sharma et al., 1998). In this context, it is worth mentioning that cyanide exists in three forms in wastewater: (1) free cyanide, such as hydrogen cyanide; (2) simple cyanide, such as sodium cyanide or potassium cyanide; and (3) complex cyanide, such as iron cyanide, nickel cyanide, and copper cyanide (Sharma et al., 1998). Further, the extensive literature review of Sharma et al. (1998) showed that other methods such as electrolytic decomposition, ozonation, electrodialysis, catalytic oxidation, reverse osmosis, ion exchange, genetic engineering applications, and photocatalyic oxidation are evolving to meet the demands of zero discharge limits. An extensive study was carried out on the degradation of CN using Fe(VI) in their single system (Castarramone et al., 2004).

Li et al. (2005) described an improved procedure for the removal of BPA (an EDC of environmental concern).

EFFECT OF FERRATE TREATMENT ON METAL IONS AND RADIONUCLIDES

Similarly, potassium ferrate can also remove a range of metals (e.g., Mn^{2+}, Cu^{2+}, Pb^{2+}, Cd^{2+}, Cr^{3+}, and Hg^{2+}) to a low level at a dose range of 10–100 mg/L as K_2FeO_4 by oxidation and coprecipitation (Jiang et al., 2006; Bartzatt et al., 1992). Arsenic(III) oxidation efficiency with ferrate(VI) was examined (Jiang et al., 2007; Fan et al., 2002). Under the given test conditions, As(III), a more toxic form, was oxidized to the less deleterious form As(V) (arsenate) by ferrate(VI) (Jiang et al., 2007; Lee et al., 2003). Likewise another preliminary work on oxidative removal of As^{3+} by ferrate(VI) was reported (Fan et al., 2002). It was observed that at a neutral pH, ferrate(VI) can oxidize As(III) in milliseconds (Lee et al., 2003). In another relevant study by Prucek et al. (2013) that used the decomposition of potassium ferrate(VI) to yield nanoparticles that have a core–shell nanoarchitecture with a γ-Fe_2O_3 core and a γ-FeOOH shell possessed with the potential to entrap arsenic. This is an alternative method using ferrate(VI) to treat As in water (Jiang, 2014). A method was investigated for simultaneously removing both heavy metals (Cu, Mn, and Zn) and natural organic matter (humic acid and fulvic acid) from river water by using potassium ferrate (K_2FeO_4), a multipurpose chemical acting as oxidant, disinfectant, and coagulant (Lim and Kim, 2010). Such efforts have been extended to the removal of chromium from alkaline high-level radioactive tank waste (Sylvester et al., 2001).

It is worth mentioning that potassium ferrate has been considered for use in acid mine drainage water treatment loaded with metallic emerging contaminants (Jiang and Lloyd, 2002; Murshed et al., 2003) and for the treatment of sulfur-containing species such as oxysulfur compounds (Murshed et al., 2003; Sharma et al., 1997) and hydrogen sulfide. Further, the chemistry of sulfide mine tailings treated with potassium ferrate (K_2FeO_4) in aqueous slurry was investigated by Murshed et al. (2003). In addition, ferrate(VI) has been used effectively to remove heavy metals such as Cr(III) (Bartzatt et al., 1992; Jiang, 2014).

Interestingly, potassium ferrate(VI) has treatability potential for certain radionuclides (Potts and Churchwell, 1994). Pioneering work based on the use of potassium ferrate to treat wastewater containing transuranic elements was performed at Los Alamos National Laboratory (Potts and Churchwell, 1994). The work demonstrated that ferrate was able to lower the gross alpha radioactivity to lower levels than an equivalent or greater amount of ferric sulfate or potassium permanganate.

EFFECT OF FERRATE TREATMENT ON PATHOGENIC MICROBES

Microbes are the most diverse group intimately linked with sewage pollution and human health. Jiang et al. (2007) demonstrated that potassium ferrate(VI) performed superior to other coagulants such as sodium hypochlorite, ferric sulfate, and aluminum sulfate in the inactivation of *Escherichia coli*. More specifically, a lower ferrate(VI) dose and a shorter contact time were required to achieve the same

E. coli-killing efficiency; the disinfection performance was less affected by the solution pH, and the disinfection rate of the ferrate(VI) was faster than that with sodium hypochlorite (Jiang, 2007). In sewage treatment, ferrate(VI) performed superiorly as an oxidant and a coagulant; it can reduce 30% more COD, while killing 3 log more bacteria.

Nevertheless, the exact mechanism of the removal of pathogenic microbes on the interface with ferrate is not clear. However, Figure 8.1 attempts to elucidate the possible mechanism as per Rai et al. (2018). Protozoan parasites such as *Cryptosporidium parvum* and *Giardia lamblia* have been known as a frequent cause of recent waterborne disease outbreaks because of their strong resistance to chlorine disinfection (Jiang, 2014). Ozone and ferrate(VI) were compared in terms of the inactivation efficiency for *Bacillus subtilis* spores, which are commonly used as an indicator of protozoan pathogens (Jiang, 2014; Makky et al., 2011).

Ferrate(VI) has been used as an oxidant in conjunction with coagulation for algal removal from lake water (Liu et al., 2009) With the combined use of ferrate(VI) and alum, algal removal increased significantly in comparison with that using alum alone (Jiang, 2014; Ma and Liu, 2002). Bacteriophage MS2, a human enteric virus detected in drinking water, is also demonstrated to be inactivated by ferrate(VI) (Hu et al., 2012; Jiang, 2014). Further, studies demonstrated that the damage of both the capsid protein and genome of MS2 increased with increasing ferrate(VI) access to the interior of the virion and therefore the extent of inactivation. Consequently, it suggests that both capsid protein and genome damage caused by the attack of ferrate(VI) may contribute to phage inactivation (Hu et al., 2012; Jiang, 2014).

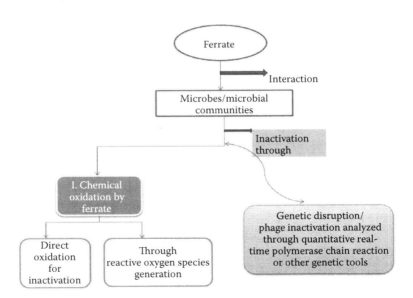

FIGURE 8.1   Possible mechanism of the action of ferrate on microbes/microbial pathogen communities.

CONCLUSION

Although the characteristic color of ferrate was first observed about 300 years ago, its chemical nature was not fully understood until the late nineteenth century. Ferrate is a multifunctional chemical reagent with great potential in water and wastewater treatment. As such, it is an environmentally friendly tool for wastewater treatment as reflected by improvements in the common indices of water pollution (such as COD, BOD, and SS.) and diverse groups of pollutants (metals, cyanides, radionuclides, EDCs, and PPCPs) and microbes inextricably linked with human health.

A stable ferrate product with high purity could be prepared by optimizing the components of raw materials and preparation conditions. At present, the most effective method for the preparation of ferrate appears to be wet oxidation. Future prospect lies in the quest for a refined preparation method applicable to a pilot-scale wastewater treatment. For instance, the *in situ* ferrate(VI) preparation technology needs to be improved further through optimization of the reactor system, reagents mass transfer, and production efficiency. Future studies should also ensure that no mutagens are produced while treating the wastewater with ferrate. It should also be noted that the performance of ferrate treatment is tightly associated with the physicochemical properties or solution conditions of water. For instance, the effect of pH on different ferrate concentrations in the degradation of micropollutants (EDCs and PPCPs) should be given more focus in future researches. Although ferrate proved to be a potent water purifier, more research studies are required in the future to validate treatment efficiency and performance determined based on laboratory studies. Moreover, the economic suitability of using ferrate(VI) needs to be assessed in various respects to provide an impetus to the drinking water industry as hazardous chemicals and microbes should be treated effectively for sustainable use in view of human health.

REFERENCES

Anquandah GAK, Sharma VK. 2009. Oxidation of octylphenol by ferrate(VI). *J Environ Sci Health A Tox Hazard Subst Environ Eng* 44:62–66.

Bartzatt R, Cano M, Johnson L, Nagel D. 1992. Removal of toxic metals and nonmetals from contaminated water. *J Toxicol Environ Health* 35(4):205–210.

Bartzatt R, Carr J. 1986. The kinetics of oxidation of simple aliphatic sulfur compounds by potassium ferrate. *Transition Met Chem* 11:116–117.

Bielski BHJ. 1991. Studies of hypervalent iron. *Free Rad Res Commun* 12–13: 469–477.

Bielski BHJ, Sharma VK, Czapski G. 1994. Reactivity of ferrate (v) with carboxylic acids: A pre-mix pulse radiolysis study. *Radiat Phys Chem* 44(5):479–484.

Carr JD, Kelter PB, Erickson AT. 1981. Ferrate(VI) oxidation of nitrilotriacetic acid. *Environ Sci Technol* 15:184–187.

Carr JD, Kelter PB, Tabatabai A, Splichal D, Erickson J, McLaughlin CW. 1995. Properties of ferrate (VI) in aqueous solution: An alternate oxidant in waste water treatment. In *Proceedings of the Conference on Water Chlorination: Environment Impact and Health Effects*, ed. RL Jolly, 1285–1298. Chelsea: MI: Lewis.

Castarramone N, Kneip A, Castetbon A. 2004. Ferrate(VI) oxidation of cyanide in water. *Environ Technol* 25:945–955.

Centi G, Perathoner S. 2003. Catalysis and sustainable (green) chemistry. *Catal Today* 77:287–297.

Ciabatti I, Tognotti F, Lombardi L. 2010. Treatment and reuse of dyeing effluents by potassium ferrate. *Desalination* 250:222–228.

Comeau F, Surette C, Brun GL, Losier R. 2008. The occurrence of acidic drugs and caffeine in sewage effluents and receiving waters from three coastal watersheds in Atlantic Canada. *Sci Total Environ* 396:132–146.

Eng YY, Sharma VK, Ray AK. 2006. Ferrate (VI): Green chemistry oxidant for degradation of cationic surfactant. *Chemosphere* 63:1785–1790.

Fan M, Brown RC, Huang CP. 2002. Preliminary studies on the oxidation of As (III) by potassium ferrate. *Int J Environ Pollut* 18:91–96.

Garcia MT, Campos E, Sanchez-Leal J, Ribosa I. 2000. Anaerobic degradation and toxicity of commercial cationic surfactants in anaerobic screening tests. *Chemosphere* 41:705–710.

García-Serna J, Pérez-Barrígon L, Cocero MJ. 2007. New trends for design towards sustain ability in chemical engineering: Green engineering. *Chem Eng J* 133:7–30.

Gilbert MB, Waite TD, Hare C. 1976. Analytical notes—An investigation of the applicability of ferrate ion for disinfection. *J Am Water Works Assoc* 68:495–497.

Gulyas H. 1997. Processes for the removal of recalcitrant organics from industrial waste waters. *Water Sci Technol* 36:9–16.

Hoppe ML, Schlemper EO, Murmann RK. 1982. Structure of dipotassium ferrate (VI). *Acta Crystallogr B* 38:2237–2239.

Hornstein BJ. 1999. Reaction mechanisms of hypervalent iron: The oxidation of amines and hydroxylamines by potassium ferrate, K_2FeO_4. PhD dissertation, New Mexico State University, Las Cruces, NM.

Horvath O, Huszank R. 2003. Degradation of surfactants by hydroxyl radicals photo generated from hydroxoiron (III) complexes: Photo chemistry. *Photochem Photobiol Sci* 2:960–966.

Hu L, Martin HM, Arce-Bulted O, Sugihara MN, Keating KA, Strathmann TJ. 2009. Oxidation of carbamazepine by Mn (VII) and Fe (VI): Reaction kinetics and mechanism. *Environ Sci Technol* 43:509–515.

Hu L, Page MA, Sigstam T, Kohn T, Mariñas BJ, Strathmann TJ, 2012. Inactivation of bacteriophage MS2 with potassium ferrate(VI). *Environ Sci Technol* 46:12079–12087.

Huang H, Sommerfeld D, Dunn B C, Lloyd CR, Eyring EM. 2001. Ferrate (VI) oxidation of aniline. *J Chem Soc. Dalton Trans* 2001:1301–1305.

Jiang JQ. 2003. Potassium ferrate (VI), a dual functional water treatment chemical. In *Proceedings of the Leading Edge Conference on Water and Wastewater Treatment Technologies*, Noordwijk/Amsterdam, The Netherlands.

Jiang JQ. 2007. Research progress in the use of ferrate (VI) for the environmental remediation. *J Hazard Mater* 146:617–623.

Jiang JQ. 2013. The role of ferrate (VI) in the remediation of emerging micro pollutants. *Procedia Environ Sci* 18:418–426.

Jiang JQ. 2014. Advances in the development and application of ferrate (VI) for water and wastewater treatment. *J Chem Tech Biotechnol* 89:165–177.

Jiang JQ, Lloyd B. 2002. Progress in the development and use of ferrate (VI) salt as an oxidant and coagulant for water and wastewater treatment. *Water Res* 36(6):1397–1408.

Jiang JQ, Panagoulopoulos A, Bauer M, Pearce P. 2006. The application of potassium ferrate for sewage treatment. *J Environ Manage* 79:215–220.

Jiang JQ, Wang S. 2003a. Enhanced coagulation with potassium ferrate (VI) for removing humic substances. *Environ Eng Sci* 20:627–635.

Jiang JQ, Wang S. 2003b. Inactivation of *Escherichia coli* with ferrate and sodium hypochlorite—A study on the disinfection performance and rate constant. In *Oxidation Technologies for Water and Wastewater Treatment*. ed., C Schroder, B Kragert, 406–411. Clausthal-Zellerfeld, Germany: Papiepflieger Verlag.

Jiang JQ, Panagoulopoulos A, Bauer M, Pearce P. 2006. The application of potassium ferrate for sewage treatment. *J Environ Manag* 79(2):215–220.

Jiang JQ, Wang S, Panagoulopoulos A. 2007. The role of potassium ferrate (VI) in the inactivation of *Escherichia coli* and in the reduction of COD for water remediation. *Desalination* 210:266–273.

Jiang JQ, Yin Q, Zhou JL, Pearce P. 2005. Occurrence and treatment trials of endocrine disrupting chemicals (EDCs) in wastewaters. *Chemosphere* 61:544–550.

Johnson MD, Hornstein BJ. 1994. Kinetics and mechanism of the ferrate oxidation of hydrazine and monomethylhydrazine. *Inorganica Chim Acta* 225:145–150.

Johnson MD, Read JF. 1996. Kinetics and mechanism of the ferrate oxidation of thiosulfate and other sulfur-containing species. *Inorg Chem* 35:6795–6799.

Johnson MD, Sharma KD. 1999. Kinetics and mechanism of the reduction of ferrate by one-electron reductants. *Inorganica Chim Acta* 293:229–233.

Kazama F. 1995. Viral inactivation by potassium ferrate. *Water Sci Technol* 31:165–168.

Khetan SK, Collins TJ. 2007. Human pharmaceuticals in the aquatic environment: A challenge to green chemistry. *Chem Rev* 107(6):2319–2364.

Kirchhoff MM. 2005. Promoting sustainability through green chemistry. *Resour Conserv Recycl* 44:237–243.

Lee C, Lee Y, Schmidt C, Yoon J, Von Gunten U. 2008. Oxidation of suspected *N*-nitrosodimethylamine (NDMA) precursors by ferrate (VI): Kinetics and effect on the NDMA formation potential of natural waters. *Water Res* 42:433–441.

Lee Y, Hwanum IK, Yoon J. 2003. Arsenic (III) oxidation by iron (VI) (ferrate) and subsequent removal of arsenic (V) by iron (III) coagulation. *Environ Sci Technol* 37:5750–5756.

Lee Y, von Gunten U. 2010. Oxidative transformation of micro pollutants during municipal wastewater treatment: Comparison of kinetic aspects of selective (chlorine, chlorine dioxide, ferrate VI and ozone) and non-selective oxidants (hydroxyl radical). *Water Res* 44:555–566.

Lee Y, Yoon J, Von Gunten U. 2005. Kinetics of the oxidation of phenols and phenolic endocrine disruptors during water treatment with ferrate (Fe(VI)). *Environ Sci Technol* 39:8978–8984.

Lee Y, Zimmermann SG, Kieu AT, von Gunten U. 2009. Ferrate (Fe(VI)) application for municipal wastewater treatment A novel process for simultaneous micropollutant oxidation and phosphate removal. *Environ Sci Technol* 43:3831–3838.

Li C, Li Z, Graham N. 2005. A study of the preparation and reactivity of potassium ferrate. *Chemosphere* 61(4):537–543.

Lim M, Kim M. 2010. Effectiveness of potassium ferrate (K_2FeO_4) for simultaneous removal of heavy metals and natural organic matters from river water. *Water Air Soil Pollut* 211:313–322.

Liu Z, Kanjo Y, Mizutani S. 2009. Removal mechanisms for endocrine disrupting compounds (EDCs) in wastewater treatment—Physical means, biodegradation, and chemical advanced oxidation: A review. *Sci Total Environ* 407(2):731–748.

Ma J, Liu W. 2002. Effectiveness and mechanism of potassium ferrate (VI) pre oxidation for algae removal by coagulation. *Water Resour* 36:871–878.

Makky EA, Park GS, Choi IW, Cho SI, Kim H. 2011. Comparison of Fe (VI) $\left(FeO_4^{2-}\right)$ and ozone in inactivating *Bacillus subtilis* spores. *Chemosphere* 83:1228–1233.

Murmann RK, Robinson PR. 1974. Experiments utilizing FeO_4^{2-} for purifying water. *Water Resour* 8:543–547.

Murshed M, Rockstraw DA, Hanson AT, Johnson M. 2003. Rapid oxidation of sulfide mine tailings by reaction with potassium ferrate. *Environ Pollut* 125:245–253.

Noorhasan N, Patel B, Sharma VK. 2010. Ferrate (VI) oxidation of glycine and glycylglycine: Kinetics and products. *Water Res* 44:927–935.

Norcross BE, Lewis WC, Gai H, Noureldin NA, Lee DG. 1997. The oxidation of secondary alcohols by potassium tetraoxoferrate (VI). *Can J Chem* 75:129–139.

Piva F, Martini L. 1998. Neurotransmitters and the control of hypophyseal gonadal functions: Possible implications of endocrine disrupters. *Pure Appl Chem* 70:1647–1656.

Potts ME, Churchwell DR. 1994. Removal of radio nuclides in wastewaters utilizing potassium ferrate (VI). *Water Environ Res* 66(2):107–109.

Prucek R, Tuček J, Kolařík J, Filip J, Marušáak Z, Sharma VK, Zbořil R. 2013. Ferrate(VI)-induced arsenite and arsenate removal by in situ structural incorporation into magnetic iron(III) oxide nanoparticles. *Environ Sci Technol* 47:3283–3292.

Rai PK. 2012. An eco-sustainable green approach for heavy metals management: Two case studies of developing industrial region. *Environ Monit Assess* 184:421–448.

Rai PK et al. 2018. A critical review of ferrate(VI)-based remediation of soil and groundwater. *J Environ Res* 160:420–448.

Read JF, Adams EK, Gass HJ, Shea SE, Theriault A. 1998b. The kinetics and mechanism of oxidation of 3-mercaptopropionic acid, 2-mercaptoethanesulfonic acid and 2-mercaptobenzoic acid by potassium ferrate. *Inorg Chim Acta* 281:43–52.

Read JF, Boucher KD, Mehlman SA, Watson KJ. 1998a. The kinetics and mechanism of the oxidation of 1,4-thioxane by potassium ferrate. *Inorganica Chim Acta* 267:159–163.

Read JF, Wyand EH.1998. The kinetics and mechanism of the oxidation of seleno-DL-methionine by potassium ferrate. *Transition Met Chem* 23:755–762.

Roche M, Toyne P. 2004. Green lead—Oxymoron or sustainable development for the lead–acid battery industry? *J Power Sources* 133:3–7.

Rush JD, Cyr JE, Zhao Z, Bielski BHJ. 1996. The oxidation of phenol by ferrate (VI) and ferrate (V): A pulse radiolysis and stopped-flow study. *Free Radical Res* 22:349–360.

Schink T, Waite TD. 1980. Inactivation of f2 virus with ferrate (VI). *Water Resour* 14:1705–1717.

Sharma VK. 2002. Potassium ferrate (VI): An environmentally friendly oxidant. *Adv Environ Res* 6:143–156.

Sharma VK. 2004. Use of Fe (VI) and Fe (V) in water and wastewater treatment. *Water Sci Technol* 49(4):69–74.

Sharma VK, Bielski BHJ. 1991. Reactivity of ferrate (VI) and ferrate (V) with amino-acids. *Inorganic Chem* 30:4306–4310.

Sharma VK, Hollyfield S. 1995. Ferrate(VI) oxidation of aniline and substituted anilines'. In *Prep. Pap. Matl. Meet.—Am. Chem. Soc. Div. Environ. Chem.*, 35:48.

Sharma VK, Luther III GW, Millero FJ. 2011. Mechanisms of oxidation of organosulfur compounds by ferrate(VI). *Chemosphere* 82:1083–1089.

Sharma VK, Mishra SK. 2006. Ferrate(VI) oxidation of ibuprofen: A kinetic study. *Environ Chem Lett* 3:182–185.

Sharma VK, Mishra SK, Nesnas N, 2006. Oxidation of sulfonamide antimicrobials by ferrate(VI) $\left[Fe(VI)O_4^{2-} \right]$. *Environ Sci Technol* 40:7222–7227.

Sharma VK, Rendon RA, Millero FJ, Vazquez FG. 2000. Oxidation of thioacetamide by ferrate(VI). *Mar Chem* 270:235–242.

Sharma VK, Rivera W, Joshi VN, Millero FJ, O'Connor D. 1999. Ferrate(VI) oxidation of thiourea. *Environ Sci Technol* 33:2645–2650.

Sharma VK, Rivera W, Smith JO, O'Brien. 1998. Ferrate (VI) oxidation of aqueous cyanide. *Environ Sci Technol* 32:2608–2613.

Sharma VK, Smith JO, Millero FJ. 1997. Ferrate (VI) oxidation of hydrogen sulfide. *Environ Sci Technol* 31:2486–2491.

Stanford C, Jiang JQ, Alsheyab M. 2010. Electrochemical production of ferrate (iron VI): Application to the wastewater treatment on a laboratory scale and comparison with iron (III) coagulant. *Water Air Soil Pollut* 209:483–488.

Sylvester P et al. 2001. Ferrate treatment for removing chromium from high-level radioactive tank waste. *Environ Sci Technol* 35:216–221.

Ternes TA. 1998. Occurrence of drugs in German sewage treatment plants and rivers. *Water Resour* 32:3245–3260.

Thorpe KL, Hutchinson TH, Hetheridge MJ, Scholze M, Sumpter JP, Tyler CR. 2001. Assessing the biological potency of binary mixtures of environmental estrogens using vitallogenin induction in juvenile rainbow trout (*Oncorhynchus mykiss*). *Environ Sci Technol* 35:2476–2481.

Warner JC, Cannon AS, Dye KM. 2004. Green chemistry. *Environ Impact Assess Rev* 24:775–799.

White DA, Franklin GS. 1998. A preliminary investigation into the use of sodium ferrate in water treatment. *Environ Technol* 19:1157–1160.

Witorsch RJ, Thomas JA. 2010. Personal care products and endocrine disruption: A critical eview of the literature. *Crit Rev Toxicol* 40:1–30.

Wood RH. 1958. The heat, free energy and entropy of the ferrate(VI) ion. *J Am Chem Soc* 80(9):2038–2041.

Yang B, Ying G, Zhao J, Zhou L, Chen F. 2012. Removal of selected endocrine disrupting chemicals (EDCs) and pharmaceuticals and personal care products (PPCPs) during ferrate (VI) treatment of secondary waste water effluents. *Water Res* 46:2194–2204.

Ying GG, Kookana RS, Kumar A. 2008. Fate of estrogens and xenoestrogens in four sewage treatment plants with different technologies. *Environ Toxicol Chem* 27(1):87–94.

Ying GG, Kookana RS, Kumar A, Mortimer M. 2009. Occurrence and implications of estrogens and xenoestrogensin sewage effluents and receiving waters from South East Queensland. *Sci Total Environ* 407(18):5147–5155.

Yngard RA, Damrongsiri S, Osathaphan K, Sharma VK. 2007. Ferrate (VI) oxidation of zinc cyanide complex. *Chemosphere* 69:729–735.

9 Phytoremediation and Nanoparticles: Global Issues, Prospects, and Opportunities of Plant–Nanoparticle Interaction in Human Welfare

INTRODUCTION

Nanotechnology is the application of science and technology to control matter at the molecular level. The expansion of nanotechnology find its multifaceted implications from genes to ecosystems. However, its basis laid by Nobel laureate Richard P. Feynman's popular lecture "There's Plenty of Room at the Bottom." A multitudinous array of nanoparticles (NPs) or nanomaterials (NMs) in concert with plants may find applications in diverse sectors of environment as well as energy, inextricably convergent with human welfare. Nanomedicine or biomedicine revolutionized the human health sector, leading to a paradigm shift in the treatment of and hope for cancer and other dreaded diseases. Nanotechnology uses NMs with at least one dimension less than 100 nm, while NMs having at least two dimensions lying between 1 and 100 nm are termed as NPs. The size of NPs is shown in Figure 9.1.

SYNTHESIS OF NPS

Top–down and bottom–up recipes are a couple of strategies for generation of nanoscale materials. It is worth mentioning at the outset that plant–NP interactions form the very basis of NP eco-synthesis, as the role of potential plants/wetland plants is emphasized in phyto/biosynthesis of NPs as living green nanofactories (bottom–up synthesis) with the in situ reduction and stabilization/capping of NPs though metabolites or biomolecules inside plant extracts. Interestingly, plant–NP interactions are the key to the green and environmentally friendly synthesis of nanoscale materials.

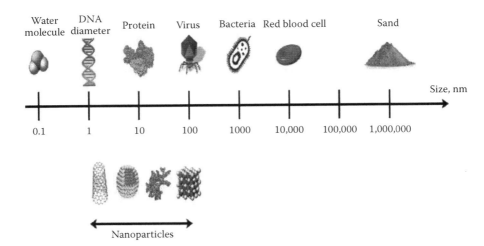

FIGURE 9.1 Size range of NPs and comparison with other features or biomolecules or microbes. (Reprinted/reproduced with permission from Prof. Visvanathan from the dissertation of M.D. Nguyen, Asian Institute of Technology, Thailand, 2013.)

RELEVANCE OF PLANT–NP INTERFACE IN MULTIFACETED ENVIRONMENTAL (GROUND WATER REMEDIATION), AGRICULTURE, AND HEALTH SECTORS

Importantly, nanoscale materials have addressed the problem of environmental contaminants in fresh, ground, or drinking water through water purification (Table 9.1); however, wastewater/sewage treatment showed limited success in NP removal. Further, the impression of plant–NP interactions on agriculture sector is of paramount importance, inextricably linked with global food security, and is adequately discussed in several researches.

To this end, plant–NP interactions have shown remarkable advances in the bioenergy sector as green and eco-sustainable energy. Plants (including microalgae and cyanobacteria; agro-ecosystem, marginal, and wetland plants), in concert with NPs, revolutionized the energy sector, specifically bioenergy/biofuel generation as a sustainable green renewable option (Figure 9.2). Interestingly, eco-technological innovations in nanoscale materials revolutionized solar, hydrogen, and bioenergy as a great input to sustainable energy.

In the environmental context, magnetic particles in particulate matter or dust may also be treated as fine particles assisting in environmental/air pollution management (Rai, 2013, 2016a–c; Rai and Panda, 2014; Rai et al., 2014, 2018; Rai and Singh, 2015; Rai and Chutia, 2016). Further, plant–NP interactions are a real hope for attaining agricultural sustainability, specifically in relation to onsite detection of pathogens, nanopesticides, nanofertilizers, and nanoherbicides. In view of the burgeoning

TABLE 9.1
Application of Nanotechnology (NPs/NMs) in Drinking Water or Ground Water Sector to Remediate Emerging Contaminants

Nanotools (NMs/NPs)	Characteristics of NMs/NPs (in Mitigating Emerging Contaminants and Eliminating Pathogenic Microbes)	Important Remarks (in Mitigating Emerging Contaminants and Eliminating Pathogenic Microbes)
Disinfection of Microbes		
Nano-Ag (nano-silver)	Affects/suppresses DNA replication and disrupts protein through Ag ion Eliminates microbes	Provides safe drinking water free from microbes
Graphene (C-based NPs) or its derivatives	Perturbs cell membrane and intracellular components, oxidative stress Eliminates microbes	Purifies water through disinfection
Carbon nanotubes (CNTs)	Membrane damage, oxidative stress to microbes like graphene-based NMs	Filters both bacteria and viruses Purifies water through disinfection
Nano-TiO$_2$	Generation of ROS eliminates microbes	Disinfects water
Fullerol and amino-fullerene	Release of ROS eliminates microbes	Water disinfection
Sensing and Monitoring		
Magnetic NPs	Super-paramagnetism enables water remediation	Water purification for drinking
Quantum dots	Wide absorption spectrum scales with the particle size and chemical component	Optical detection and purification
CNTs	Large surface area and high mechanical strength and chemical stability	Electrochemical detection, sample preconcentration
Photocatalysis		
Fullerene-based NMs	Solar photocatalysis and selectivity	Wastewater purification through solar disinfection
Membrane Nanotechnologies		
Nanomagnetite	Super-paramagnetic in nature	Purify drinking water through forward osmosis
CNTs	Disinfection as listed earlier	Filters both bacteria and viruses
Nano-zeolites	Act as molecular sieve	High permeability thin-film nanocomposite membranes
Nano-Ag	Antimicrobial in nature	Used as anti-biofouling membranes

(Continued)

TABLE 9.1 (CONTINUED)
Application of Nanotechnology (NPs/NMs) in Drinking Water or Ground Water Sector to Remediate Emerging Contaminants

Nanotools (NMs/NPs)	Characteristics of NMs/NPs (in Mitigating Emerging Contaminants and Eliminating Pathogenic Microbes)	Important Remarks (in Mitigating Emerging Contaminants and Eliminating Pathogenic Microbes)
	Adsorption	
CNTs	Wide surface area; more adsorption sites for environmental contaminants; ease in reuse	Detection, adsorption of recalcitrant/ non-biodegradable contaminants
Nanoscale metal oxide	High surface area, more adsorption sites for environmental contaminants; ease in reuse, like CNTs	Used as adsorptive media filters, slurry reactors to purify water

human population, food security is a very serious issue faced by agriculture scientists. Nanorevolution should have the potential to cause a paradigm shift in agriculture sustainability besides addressing the food security challenges and keeping the environmental values intact.

Excessive use of NPs can release them into ambient air, an integral component of the environment besides surface and ground water. Air pollution is a global challenge faced by both developing and developed nations, and release of NPs in air exacerbates this problem in addition to other emerging air contaminants. Metallic NPs in the atmosphere may be a matter of concern in view of human health perspective; thus, study of their fate in air is extremely important (Rai, 2013, 2016a–c; Rai et al., 2014; Rai and Panda, 2014; Rai and Singh, 2015; Rai and Chutia, 2016).

In earlier chapters, we discussed the phytotechnologies used for the removal of emerging contaminants with great documentation in the literature (Rai, 2007, 2008a–c, 2009, 2010a–d, 2011, 2012; Singh and Rai, 2016). In the context of this chapter, and in continuation with our opportunities component of SWOT analysis (in Chapter 3), the author envisages that coupling of phytotechnologies as well as nanotechnologies with bioenergy/biofuel production using plants/wetland plants from marginal or poorly arable lands or contaminated wetland systems/ aquatics may expand the ecological footprint (land required for sustaining natural resources) and reduce carbon footprint (CO_2 emission at various levels, contributing to climate change). Plant–NP interfaces are extremely vital in boosting a plethora of energy and environmental sectors in an efficient, green, and economic manner (Figure 9.2).

After giving a concise overview of the relevance of this chapter, I will confine my quest to phytotechnologies and plant–NP interactions in environmental amelioration. Interface with plants and NPs can be extremely relevant in the context of aquatic/wetland environment.

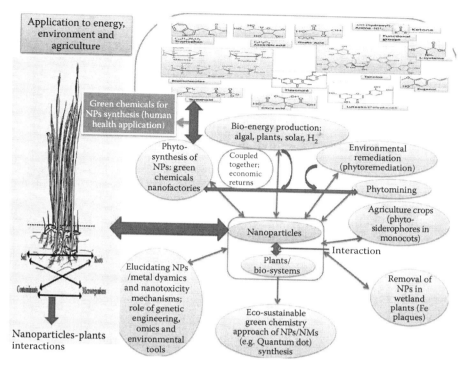

FIGURE 9.2 Broad spectrum of applications as an outcome of plant–NP interfaces.

WETLAND PLANT–NP INTERFACES: IMPLICATIONS FOR PHYTOREMEDIATION AND PHYTOTECHNOLOGIES

Algae form the base of the food web; however, some algae and cyanobacteria result in algal bloom, causing confrontational impact on the environment and drinking water through odor problems (Lee et al., 2017). Diverse NPs proved their worth in addressing this problem in aquatics and further assisted in drinking water purification, intimately linked with human health.

Aquatic/wetland plant–NP interaction can transmogrify different environmental technologies inextricably linked with environmental remediation (phytoremediation, phytomining, phytosynthesis of NPs, and their mechanisms) besides implications in bioenergy sector (Figure 9.2).

The mechanisms involved in phytoremediation of emerging contaminants are elucidated at the molecular level; however, little research exists in unravelling the NP phytoremediation pathway. Phytoremediation in wetland/aquatic plants or macrophytes involve plant–soil–microbe interactions, cell wall binding, cytoplasmic chelators, symplastic loading, ion exchange, etc., as demonstrated in earlier chapters (Chapters 2 and 3) for metallic and other emerging contaminants (Rai, 2007, 2008a–c, 2009, 2010a–d, 2012; Singh and Rai, 2016). To this end, several instruments like nano-Secondary Ion Mass Spectrometry (SIMS), particle-induced x-ray emission, and x-ray spectroscopy devices mentioned in a couple of figures in Chapter 5 proved useful with certain limitations. Recently, a novel instrument, synchrotron x-ray nanofluorescence, was used in the root system of a wetland plant/ macrophyte *Spartina alterniflora* (during dormancy) to study the distribution of metals like As (metalloid), Ca, Cr, Cu, Fe, Mn, Ni, S, and Zn, and results revealed that a casparian strip lying in the outer endodermis of the *Spartina* root played a significant role in metal/metalloid uptake (Feng et al., 2017).

NPs may discover their way into the aquatic environment and natural and constructed wetlands (CWs) through the mechanism of sorption. In this context, wetland substrates and biofilm assisted in AgNP removal from CWs (Auvinen et al., 2016). To this end, several wetland plants/macrophytes act as a model (due to frequent abundance, rapid growth, and thrust for emerging environmental contaminants, as discussed in Chapter 3) for ecotoxicity resulting from NPs, like members of the duckweed family (specifically *Lemna* and *Spirodela* for AgNPs) (Kim et al., 2011). Further, duckweed, *Landoltia punctata* confirmed phytoaccumulation of AgNPs, Ag_2S NPs, and $AgNO_3$ on their interaction, and compartmentalization in plants was done through instruments X-ray fluorescence (XRF) and Extended X-Ray Absorption Fine Structure (EXAFS) (see figure of Chapter 5 for details on instrumentation). Results of the research further reflected that AgNPs had transmuted itself into silver sulfide and silver thiol species (Stegemeier et al., 2017).

A growing body of research demonstrated that wetland plants are extremely competent tools that can be used in the phytoremediation of emerging environmental contaminants/NPs having human health implications (Table 9.2). The quest for the fate of AgNPs (Polyvinylpyrrolidone [PVP] coated) in an emergent wetland with certain wetland plants like *Juncus effuses, Carex lurida, Panicum virgatum*, and *Lobelia cardinalis* demonstrated that AgNPs were confined to belowground portion (root) and a very few amount translocate to aboveground biomass (shoot) (Lowry et al., 2012). Likewise, CuO- and ZnO-NPs also phytoremediated in the root zone of a couple of potent wetland plants, *Phragmites australis* and *Schoenoplectus tabernaemontani* (Zhang et al., 2014, 2015).

To this end, citrate-coated AgNP phytoremediation in a hydroponic culture of a wetland plant, *Elodea canadensis*, has been demonstrated up to the range of 57% (Van Koetsem et al., 2015). Interestingly, surface coating and tailoring of NPs/NMs may affect their phytoremediation in aquatic environment. Further, molecular and instrumental tools are being used to investigate the interactions between phytoremediation and NPs. To this end, synchrotron microanalyses elucidated the interface mechanisms (between NPs and wetland plants) through which wetlands plants/macrophytes like *P. australis* and *Iris pseudacorus* can transform copper into metallic CuNPs in belowground biomass (root) lying in contaminated soil with the assistance of endomycorrhizal fungi. Cu/ NP phytoremediation can be mediated through their binding to thiol (-SH) groups and phytochelatins (cysteine-rich metal binding peptides) (Manceau et al., 2008).

In this context, an important research investigation revealed that Fe nanocomplexes/ plaques formed in 16 plants were localized in the root (ca. 97%), even in sterile conditions with complete absence of microbial associations, which are assumed to enhance the uptake of metals or metal bound complexes. Further, the dominance of Fe nanocomplexes/plaques in the root may be considered as a homeostatic

TABLE 9.2

Phytoremediation of NPs through Wetland Plants/Macrophytes: Eco-Technological and Nanoecotoxicology Implications

S.No.	Diverse NPs	Wetland Plants	Implications/Effect on Wetland Plants	References
1	Ag (citrate coated)	E. canadensis	Hydroponic culture of a wetland plant E. canadensis has demonstrated up to 57% phytoremediation of citrate-coated AgNPs	Van Koetsem et al., 2015
2	Ag (PVP coated)	J. effuses, C. lurida, L. cardinalis, and P. virgatum	AgNP (PVP coated) phytoremediation with these plants in an emergent wetland was confined to belowground portion (root) and very few translocate to aboveground biomass (shoot)	Lowry et al., 2012
3	CuO- and ZnO-NPs	P. australis and S. tabernaemontani	CuO- and ZnO-NP phytoremediation mainly through roots (rhizoremediation)	Zhang et al., 2014, 2015
4	AgNPs	Lemna minor (duckweed)	AgNPs (20–100 nm with conc. 5 μg g⁻¹) decreased the growth of this wetland plant in a size-dependent manner with phytoremediation implications	Gubbins et al., 2011
5	FeNPs	Eichhornia crassipes (water hyacinth)	FeNP phytoremediation through E. crassipes was coupled with bioenergy production; E. crassipes may act as a wetland plant model biosystem to elucidate NP uptake	Mahmood et al., 2012; Rani et al., 2016
6	AgNPs	P. stratiotes	Phytoremediation AgNP and Ag ions under 0.02 mg L⁻¹ concentration	Hans et al., 2015
7	Ag, Ti, and Zn NPs	L. minor and Spirodela polyrhiza (duckweed)	Used in NP phytoremediation due to frequent abundance, rapid growth, and thrust for emerging environmental contaminants	Kim et al., 2011; Farrag, 2015
8	Metallic NPs-Mn, Cu, Zn, Ag + Ag₂O	E. canadensis, N. guadelupensis, V. spiralis, R. fluitans (submerged) and P. stratiotes, S. natans, and L. laevigatum (free-floating wetland plants)	Phytoremediation of metallic NPs ranged from 30%–100% with these wetland plants (free floating and submerged—together termed "pleuston"). P. stratiotes and S. natans were the most potent wetland plants in NP phytoremediation	Olkhovych et al., 2016
9	PVP-AgNP and GA-AgNP	Diverse wetland plants	Phytoremediation was not sustained due to reduction in growth at 1, 10, or 40 mg L⁻¹	Yin et al., 2012

(Continued)

TABLE 9.2 (CONTINUED)

Phytoremediation of NPs through Wetland Plants/Macrophytes: Eco-Technological and Nanoecotoxicology Implications

S.No.	Diverse NPs	Wetland Plants	Implications/Effect on Wetland Plants	References
10	ZnO NPs	*Prosopis farcta*	Phytoremediation of ZnO NPs at various concentrations	Yahya and Al-Salih, 2014
11	Alumina (Al_2O_3) NP	*L. minor* (duckweed)	Escalated *L. minor* growth with increased tendency for biomass accumulation having implications for alumina NP phytoremediation	Juhel et al., 2011
12	Ti or T-oxide NPs	*Rumex crispus* and *Elodea canadian*	Pronounced phytoremediation of TiNP in these wetland plants in conjunction with couple of agro-crops (*Phaseolus vulgaris* and *Triticum aestivum*) at 0–30 mg L^{-1} and 0–20 mg L^{-1} for *E. canadian*	Jacob et al., 2013
13	CuNPs	*P. australis* and *I. pseudacorus*	Phytoaccumulation at the soil–root interface with association of endomycorrhizal fungi initially as Cu and later as CuNPs; potential instrumentation also assisted this interaction and concomitantly elucidating nanotoxicity mechanisms through Synchotron-based microanalytical and imaging tools	Manceau et al., 2008
14	AgNPs and $AgNO_3$	*S. polyrhiza* (duckweed); used as a model plant to study AgNP phytoaccumulation	Phytoremediation of AgNPs followed by biochemical responses, root abscission, and disintegration of duckweed colonies; $AgNO_3$ proved more toxic than AgNPs	Jiang et al., 2012
15	Ag ion	*Potamogeton crispus*	Phytoaccumulation-induced toxicity	Xu et al., 2010
16	AuNPs	*Ceratophyllum demersum*	Capable of binding AuNPs and further studies are needed in phytoremediation context	Ostrumov and Kolesov, 2010
17	AuNPs	*S. alterniflora*	Capable of binding/immobilization of AuNPs with phytoremediation implications	Ferry et al., 2009

(Continued)

TABLE 9.2 (CONTINUED)
Phytoremediation of NPs through Wetland Plants/Macrophytes: Eco-Technological and Nanoecotoxicology Implications

S.No.	Diverse NPs	Wetland Plants	Implications/Effect on Wetland Plants	References
18	CuO NPs	*Elodea canadensis*	Phytoremediation of CuO NPs in microcosms with significant accumulation of Cu in biomass; *Elodea* biomass is able to immobilize/bind to CuO NPs with some toxicity	Johnson et al., 2011
19	AgNPs	*Chlamydomonas reinhardtii* wetland green algae	Nanoecotoxicology investigations revealed release of Ag ions on interaction of AgNPs under the influence of several biotic factors	Navarro et al., 2008
20	AgNPs	*P. australis*	Phytoremediation of AgNPs with this wetland plant was more efficient than Ag ions in salt marshes; however, the ecosystem dynamics of wetland microbes or rhizo-sediments may be affecting this green technology	Fernandes et al., 2017
21	CuONPs	*Halimione portulacoides* and *P. australis*	These salt marsh wetland plants showed considerable phytoremediation (rhizoremediation) of CuONPs; however, rhizo-sediment's microbial community structure was affected differentially. CuONPs decreased microbial community in *H. portulacoides*, while microbial diversity was increased in *P. australis*	Fernandes et al., 2017
22	TiO₂ NPs	*S. polyrhiza* (duckweed)	Phytoremediation process of TiO₂ NPs with this duckweed witnessed remarkable changes in biochemical parameters, increased SOD to combat ROS, and decreased peroxidase to boost defense mechanisms in wetland plant	Movafeghi et al., 2016
23	Maghemite NPs	*P. australis*	Maghemite NPs encourage Fe plaque formation in this wetland plant, which entraps and and leads to arsenic phytoremediation	Pardo et al., 2017

mechanism to protect the aboveground biomass from toxicity of nanocomplexes (Pardha-Saradhi et al., 2014). Also, nanotoxicity studies in wetland plants, as demonstrated in *Eichhornia crassipes*, may elucidate the phytoremediation mechanisms of NPs (Rani et al., 2016).

Wetland plants/macrophytes, both free floating and submerged (together termed as "pleuston"), have an immense potential for phytoremediation of metallic NPs, like Mn, Cu, Zn, and Ag + Ag_2O lying in aquatic environment. To this end, submerged wetland plants (*E. canadensis, Najas guadelupensis, Vallisneria spiralis*, and *Riccia fluitans*) and free-floating hydrophytes (*Limnobium laevigatum, Pistia stratiotes*, and *Salvinia natans*) efficiently resulted in phytoremediation (ranging from 30% to 100%) of the aforesaid metal NPs existing in colloidal systems and among seven wetland plants, with *P. stratiotes* and *S. natans* being the most potent phytoremediation tools in view of their photosynthetic system stability during NP action, and deserve incorporation in future pilot-scale NP eliminating biosystems. In this research, photosynthetic pigments change drastically in submerged wetland plants, unlike floating ones (Olkhovych et al., 2016). A rooted wetland plant, *E. crassipes*, removed metals (lead, Pb; copper, Cu; zinc, Zn; chromium, Cr; and cadmium, Cd) and was able to extract metals as NPs (Mahmood et al., 2012). Further, the effects of alumina NPs on the morphological as well as physiological attributes of *Lemna minor* revealed interesting observations that alumina NPs markedly increased biomass accumulation of *L. minor* due to increased efficiencies in the light reactions of photosynthesis (Juhel et al., 2011). Moreover, a couple of wetland plants (*Rumex crispus* and *E. canadensis*) were able to phytoremediate TiNPs along with two plants from agroecosystems. The bioavailability of nanoscale metal contaminants like TiO_2 NPs remarkably affects phytoaccumulation as well as subsequent phytoremediation, in addition to environmental impacts.

Interestingly, synthetic NM (nanomaghemite, nFe_2O_3) and $FeSO_4$, in conjunction, encourage Fe plaque formation (also discussed in earlier chapters) in the wetland plant *P. australis*, which in turn assists in arsenic entrapment (Pardo et al., 2016). Further, the phytoremediation potential of *P. stratiotes* in AgNP and Ag ion contaminated wastewaters was tested, and the results revealed that this wetland plant can survive in AgNP and ions under 0.02 mg L^{-1} while the contaminants find their way inside the plant bio-system and thus may be used in NP phytoremediation (Hans et al., 2015).

RECENT ADVANCES AND FUTURE PROSPECTS OF NP–PHYTOREMEDIATION INTERFACE

Recent advances in this context demonstrated that sensor technologies also found a cutting-edge place for in situ phytoremediation sector as a disposable bismuth film electrode assessed the efficiency of *L. minor* in Pb and Cd removal (Neagu et al., 2014). The biodiversity of several terrestrial (including plants of agriculture systems) and aquatic plant species, specifically wetland plants/macrophytes as well as algae, has been recognized, which can be used for simultaneous phytoremediation as well as procurement of beneficial end products such as bioethanol, biodiesel, wood, charcoal, bioplastics, etc., through functioning of green phyto-technologies (Table 9.2).

We found that concurrent phytoremediation and bioenergy asset generation corroborated the need for growing bioenergy plants. To this end, phycoremediation (algal bioremediation, as discussed in an earlier chapter) and phytoremediation may be coupled with bioenergy production as a sustainable energy source, besides, eco-removing the emerging contaminants of concern (Rai, 2008b, 2009a, 2011, 2012).

More researches are warranted in addressing the biomass-based energy issue in order to establish an interface between NPs and phytoremediation. Moreover, the impacts of NPs on plant–soil–microbe interactions are extremely relevant in an environmental perspective and more focused researches can extend our understanding on plant–NP interactions. Further, extensive research is required on stress tolerance mechanisms in wetland plants resulting from NP interactions.

REFERENCES

Auvinen H, Sepúlveda VV, Rousseau DPL, Laing GD. 2016. Substrate- and plant-mediated removal of citrate-coated silver nanoparticles in constructed wetlands. *Environ Sci Pollut Res* 23:21920–21926.

Farrag HF. 2015. Evaluation of the growth response of *Lemna gibba* L. (duckweed) exposed to silver and zinc oxide nanoparticle. *World Appl Sci J* 33(2):190–202.

Feng H et al. 2017. Nanoscale measurement of trace element distributions in *Spartina alterniflora* root tissue during dormancy. *Sci Rep* 7:40420.

Fernandes JP, Mucha AP, Francisco T, Gomes CR, Almeida CMR. 2017. Silver nanoparticles uptake by salt marsh plant—Implications for phytoremediation processes and effects in microbial community dynamics. *Mar Pollut Bull* 176–183.

Ferry JL et al. 2009. Transfer of gold nanoparticles from the water column to the estuarine food web. *Nat Nanotechnol* 4:441.

Gubbins EJ, Batty LC, Lead JR. 2011. Phytotoxicity of silver nanoparticles to *Lemna minor* L. *Environ Pollut* 59:1551–1559.

Hans NA, Caruso JA, Zhang P. 2015. Assessing *Pistia stratiotes* for phytoremediation of silver nanoparticles and Ag(I) contaminated waters. *J Environ Manage* 164:41–45.

Jacob DL et al. 2013. Uptake and translocation of Ti from nanoparticles in crops and wetland plants. *Int J Phytoremediation* 15:142–153.

Jiang HS, Li M, Chang F, Li W, Yin L. 2012. Physiological analysis of silver nanoparticles and AgNO3 toxicity to *Spirodela polyrhiza*. *Environ Toxicol Chem* 31(8):1880–1886.

Johnson ME et al. 2011. Study of the interactions between *Elodea canadensis* and CuO nanoparticles. *Russian J Gen Chem* 81(13):2688–2693.

Juhel G et al. 2011. Alumina nanoparticles enhance growth of *Lemna minor*. *Aquat Toxicol* 105:328–336.

Kim E, Kim S, Kim H, Lee S, Lee S, Jeong S. 2011. Growth inhibition of aquatic plant caused by silver and titanium oxide nanoparticles. *Toxicol Environ Health Sci* 3(1):1–6.

Lee J, Rai PK, Jeon YJ, Kim KH, Kwon EE. 2017. The role of algae and cyanobacteria in the production and release of odorants in water. *Environ Pollut* 227:252–262.

Lowry GV et al. 2012. Long-term transformation and fate of manufactured Ag nanoparticles in a simulated large scale freshwater emergent wetland. *Environ Sci Technol* 46:7027–7036.

Mahmood T et al. 2012. Metallic phytoremediation and extraction of nanoparticles *Int J Phys Sci* 7(46):6105–6116.

Manceau A et al. 2008. Formation of metallic copper nanoparticles at the soil–root interface. *Environ Sci Technol* 42:1766–1772.

Movafeghi A, Khataee AR. Moradi Z, Vafaei F. 2016. Biodegradation of direct blue 129 diazo dye by *Spirodela polyrrhiza*: An artificial neural networks modeling. *Int J Phytoremediation* 18(4):337–347.

Navarro E, Piccipetra F, Wagner B, Marconi F, Kaegi R, Odzak N, Sigg L, Behra R. 2008. Toxicity of silver nanoparticles to *Chlamydomonas reinhardtii*. *Environ Sci Technol* 42:8959–8964.

Neagu D et al. 2014. Disposable electrochemical sensor to evaluate the phytoremediation of the aquatic plant *Lemna minor* L. toward Pb(2+) and/or Cd(2+). *Environ Sci Technol* 48:7477–7485.

Olkhovych O et al. 2016. Removal of metal nanoparticles colloidal solutions by water plants. *Nanoscale Res Lett* 11:518.

Ostrumov SA, Kolesov GM. 2010. The aquatic macrophyte *Ceratophyllum demersum* immobilizes Au nanoparticles after their addition to water. *Dokl Biol Sci* 431:124.

Pardha-Saradhi P, Yamal G, Peddisetty T, Sharmila P, Singh J, Nagarajan R, Rao KS. 2014. Plants fabricate Fe-nanocomplexes at root surface to counter and phytostabilize excess ionic Fe. *Biometals* 27:97–114.

Pardo T et al. 2016. Maghemite nanoparticles and ferrous sulfate for the stimulation of iron plaque formation and arsenic immobilization in *Phragmites australis*. *Environ Pollut* 219:296–304.

Rai PK. 2007. Wastewater management through biomass of *Azolla pinnata*: An ecosustainable approach. *Ambio* 36(5):426–428.

Rai PK. 2008a. Phytoremediation of Hg and Cd from industrial effluents using an aquatic free floating macrophyte *Azolla pinnata*. *Int J Phytoremediation* 10(5):430–439.

Rai PK. 2008b. Heavy-metal pollution in aquatic ecosystems and its phytoremediation using wetland plants: An ecosustainable approach. *Int J Phytoremediation* 10(2):133–160.

Rai PK. 2008c. Mercury pollution from chlor-alkali industry in a tropical lake and its biomagnification in aquatic biota: Link between chemical pollution, biomarkers and human health concern. *Hum Ecol Risk Assess Int J* 14:1318–1329.

Rai PK. 2009. Heavy metal phytoremediation from aquatic ecosystems with special reference to macrophytes. *Crit Rev Environ Sci Technol* 39(9):697–753.

Rai PK. 2010a. Microcosm investigation on Phytoremediation of Cr using *Azolla pinnata*. *Int J Phytoremediation* 12:96–104.

Rai PK. 2010b. Phytoremediation of heavy metals in a tropical impoundment of industrial region. *Environ Monit Assess* 165:529–537.

Rai PK. 2010c. Seasonal monitoring of heavy metals and physico-chemical characteristics in a Lentic ecosystem of sub-tropical industrial region, India, *Environ Monit Assess* 165:407–433.

Rai PK. 2010d. Heavy metal pollution in Lentic ecosystem of sub-tropical industrial region and its phytoremediation. *Int J Phytoremediation* 12(3):226–242.

Rai PK. 2011. *Heavy Metal Pollution and Its Phytoremediation though Wetland Plants*. New York: Nova Science Publisher, pp. 196. ISBN no. 978-1-61209-938-5.

Rai PK. 2012. An Eco-sustainable Green Approach for Heavy metals Management: Two Case Studies of Developing Industrial Region. *Environ Monit Assess* 184:421–448.

Rai PK. 2013. Environmental magnetic studies of particulates with special reference to biomagnetic monitoring using roadside plant leaves. *Atmos Environ* 72:113–129.

Rai PK. 2016a. Biodiversity of roadside plants and their response to air pollution in an Indo-Burma hotspot region: Implications for urban ecosystem restoration. *J Asia Pac Biodivers* 9:47–55.

Rai PK. 2016b. *Biomagnetic Monitoring through Roadside Plants of an Indo-Burma Hot Spot Region*. UK: Elsevier, pp. 198.

Rai PK. 2016c. Impacts of particulate matter pollution on plants: Implications for environmental biomonitoring. *Ecotoxicol Environ Saf* 129:120–136.

Rai PK, Chutia BM, Patil SK. 2014. Monitoring of spatial variations of particulate matter (PM) pollution through bio-magnetic aspects of roadside plant leaves in an Indo-Burma hot spot region. *Urban For Urban Gree* 13:761–770.

Rai PK, Chutia BM. 2016. Particulate matter bio-monitoring through magnetic properties of an Indo-Burma hotspot region. *Chem Ecol* 32(6):550–574.

Rai PK, Panda LS. 2014. Dust capturing potential and air pollution tolerance index (APTI) of some roadside tree vegetation in Aizawl, Mizoram, India: An Indo-Burma hot spot region. *Air Qual Atmos Health* 7(1):93–101.

Rai PK, Singh MM. 2015. *Lantana camara* invasion in urban forests of an Indo-Burma hotspot region and its ecosustainable management implication through biomonitoring of particulate matter. *J Asia Pac Biodivers* 8:375–381.

Rai PK et al. 2018. A critical review of ferrate(VI)-based remediation of soil and groundwater. *J Environ Res* 160:420–448.

Rani PU et al. 2016. Effect of synthetic and biosynthesized silver nanoparticles on growth, physiology and oxidative stress of water hyacinth: *Eichhornia crassipes* (Mart) Solms. *Acta Physiol Plant* 38:1–9.

Singh MM, Rai PK. 2016. Microcosm investigation of Fe (iron) removal using macrophytes of Ramsar lake: A phytoremediation approach; *Int J Phytoremediation* 18(12):1231–1236.

Stegemeier JP, Colman BP, Schwab F, Wiesner MR, Lowry GV. 2017. Uptake and distribution of silver in the aquatic plant *Landoltia punctata* (Duckweed) exposed to silver and silver sulfide nanoparticles. *Environ Sci Technol* 51:4936–4943.

Van Koetsem F et al. 2015. Fate of engineered nanomaterials in surface water: Factors affecting interactions of Ag and CeO_2 nanoparticles with (re)suspended sediments. *Ecol Eng* 80:140–150.

Xu S, Hu JZ, Xie KB, Yang HY, Du KH, Shi GX. 2010. Accumulation and acute toxicity of silver in *Potamogeton crispus* L. *J Hazard Mater* 173:186–193.

Yahya RT, Al-Salih HS. 2014. Uptake of Zinc Nanoparticles by Prosopisfarcta L. Plants Callus Cultures. *Eng Technol J* 32(3):615–621.

Yin L, Colman BP, McGill BM, Wright JP, Bernhardt ES. 2012. Effects of silver nanoparticle exposure on germination and early growth of eleven wetland plants *PLoS One* 7(10):1–7.

Zhang D et al. 2014. Uptake and accumulation of CuO nanoparticles and CdS/ZnS quantum dot nanoparticles by *Schoenoplectus tabernaemontani* in hydroponic mesocosms. *Ecol Eng.* 70:114–123.

Zhang D et al. 2015. Phytotoxity and bioaccumulation of ZnO nanoparticles in *Schoenoplectus tabernaemontani*. *Chemosphere* 120:211–219.

Appendix

TABLE A.1
Correlation between Physicochemical Parameters of Site I

Site I

	Temperature	pH	Transparency	TS	DO	BOD	Acidity	Alkalinity	Chloride	Total Hardness	Turbidity	Nitrate	Phosphate
Temperature	1.00												
pH	−0.45	1.00											
Transparency	−0.61	0.35	1.00										
TS	0.69	−0.91	−0.68	1.00									
DO	−0.75	−0.13	0.40	−0.13	1.00								
BOD	−0.21	−0.44	0.42	0.20	0.72	1.00							
Acidity	0.96	−0.20	−0.54	0.49	−0.84	−0.31	1.00						
Alkalinity	−0.81	0.82	0.60	−0.93	0.25	−0.24	−0.65	1.00					
Chloride	−0.97	0.65	0.65	−0.84	0.60	0.09	−0.87	0.91	1.00				
Total hardness	−0.76	0.87	0.53	−0.94	0.18	−0.33	−0.58	0.99	0.88	1.00			
Turbidity	0.83	0.03	−0.33	0.22	−0.96	−0.52	0.92	−0.35	−0.69	−0.29	1.00		
Nitrate	−0.27	0.95	0.35	−0.85	−0.36	−0.54	−0.01	0.76	0.49	0.81	0.28	1.00	
Phosphate	−0.12	0.73	0.13	−0.56	−0.38	−0.56	0.11	0.62	0.28	0.67	0.39	0.85	1.00

Note: TS, total solids; DO, dissolved oxygen; BOD, biological oxygen demand.

TABLE A.2
Correlation between Physicochemical Parameters of Site II

Site II

	Temperature	pH	Transparency	TS	DO	BOD	Acidity	Alkalinity	Chloride	Total Hardness	Turbidity	Nitrate	Phosphate
Temperature	1.00												
pH	0.47	1.00											
Transparency	-0.49	-0.89	1.00										
TS	0.25	-0.37	0.44	1.00									
DO	-0.68	-0.85	0.90	0.43	1.00								
BOD	0.09	0.09	0.67	0.81	0.66	1.00							
Acidity	0.95	0.68	-0.74	-0.03	-0.87	-0.21	1.00						
Alkalinity	-0.77	0.17	-0.08	-0.67	0.08	-0.64	-0.54	1.00					
Chloride	0.67	0.71	-0.84	-0.41	-0.83	-0.37	0.85	-0.23	1.00				
Total hardness	0.16	0.83	-0.87	-0.76	-0.75	-0.77	0.47	0.42	0.78	1.00			
Turbidity	0.72	0.88	-0.83	-0.35	-0.97	-0.58	0.88	-0.11	0.78	0.69	1.00		
Nitrate	-0.03	0.81	-0.71	-0.80	-0.66	-0.87	0.27	0.66	0.50	0.91	0.65	1.00	
Phosphate	0.02	0.82	-0.77	-0.81	-0.69	-0.84	0.33	0.59	0.61	0.96	0.66	0.99	1

Note: TS, total solids; DO, dissolved oxygen; BOD, biological oxygen demand.

TABLE A.3
Correlation between Physicochemical Parameters of Site III

Site III

	Temperature	pH	Transparency	TS	DO	BOD	Acidity	Alkalinity	Chloride	Total Hardness	Turbidity	Nitrate	Phosphate
Temperature	1.00												
pH	-0.58	1.00											
Transparency	-0.36	-0.46	1.00										
TS	0.71	-0.77	0.30	1.00									
DO	-0.62	-0.08	0.90	0.09	1.00								
BOD	0.44	-0.31	-0.10	0.50	-0.22	1.00							
Acidity	0.98	-0.57	-0.42	0.67	-0.66	0.50	1.00						
Alkalinity	-0.87	0.80	-0.13	-0.95	0.16	-0.49	-0.82	1.00					
Chloride	-0.93	0.78	0.02	-0.86	0.32	-0.33	-0.88	0.96	1.00				
Total hardness	-0.94	0.77	0.04	-0.88	0.35	-0.52	-0.90	0.98	0.98	1.00			
Turbidity	0.56	0.22	-0.94	-0.16	-0.98	0.16	0.58	-0.09	-0.24	-0.27	1.00		
Nitrate	-0.75	0.88	-0.34	-0.96	-0.04	-0.35	-0.69	0.97	0.93	0.91	0.12	1.00	
Phosphate	-0.63	0.87	-0.33	-0.89	-0.05	-0.70	-0.66	0.87	0.76	0.17	0.17	0.87	1.00

Note: TS, total solids; DO, dissolved oxygen; BOD, biological oxygen demand.

TABLE A.4
Correlation between Physicochemical Parameters of Site IV

Site IV

	Temperature	pH	Transparency	TS	DO	BOD	Acidity	Alkalinity	Chloride	Total Hardness	Turbidity	Nitrate	Phosphate
Temperature	1.00												
pH	0.21	1.00											
Transparency	-0.60	-0.89	1.00										
TS	0.33	-0.76	0.47	1.00									
DO	-0.88	-0.23	0.59	-0.06	1.00								
BOD	-0.39	-0.24	0.35	0.22	0.53	1.00							
Acidity	0.96	0.18	-0.59	0.26	-0.95	-0.36	1.00						
Alkalinity	-0.69	0.34	0.08	-0.73	0.55	-0.14	-0.74	1.00					
Chloride	0.85	0.07	-0.45	0.41	-0.75	0.09	0.89	-0.85	1.00				
Total hardness	-0.75	0.43	0.01	-0.78	0.62	0.23	-0.75	0.90	-0.70	1.00			
Turbidity	0.51	0.72	-0.76	-0.53	-0.61	-0.74	0.45	0.25	0.15	0.07	1.00		
Nitrate	0.27	0.83	-0.73	-0.73	-0.39	-0.51	0.22	0.47	0.01	0.38	0.94	1.00	
Phosphate	0.27	0.81	-0.75	-0.79	-0.53	-0.50	0.33	0.35	0.10	0.32	0.93	0.93	1.00

Note: TS, total solids; DO, dissolved oxygen; BOD, biological oxygen demand.

TABLE A.5
Permissible Limits of Physicochemical Parameters of Water by Different Scientific Agencies

Parameter	Standards			
	USPH	ISI	WHO	ICMR
Temperature (°C)	–	40	–	–
pH (nM/L)	6–8.5	6–9	6.5–8.5	7–8.5
TS (mg/L)	–	–	500	500–1500
Turbidity (NTU)	–	–	5	–
Total hardness (mg/L $CaCO_3$)	500	–	–	300
DO (mg/L)	>4	>5	–	–
BOD (mg/L)	–	<3	–	–
Chloride (mg/L $CaCO_3$)	250	600	200	250
Total alkalinity (mg/L $CaCO_3$)	–	200	–	–
Nitrate (mg/L)	10	50	–	20
Phosphate (mg/L)	0.1	–	–	–
Fe (in water)	–	–	1 mg/L	–
Fe (in biomass)	–	–	20 mg/kg	–
Zn (in water)	–	–	5 mg/L	–
Zn (in biomass)	–	–	50 mg/kg	–
Pb (in water)	–	–	0.05 mg/L	–
Pb (in biomass)	–	–	2 mg/kg	–
Cd (in water)	–	–	0.01 mg/L	–
Cd (in biomass)	–	–	0.02 mg/kg	–
Cr (in water)	–	–	0.1 mg/L	–
Cr (in biomass)	–	–	1.3 mg/kg	–
Hg (in water)	–	–	0.001 mg/L	–
Hg (in biomass)	–	–	–	–
As (in water)	–	–	0.05 mg/L	–
As (in biomass)	–	–	–	–

Note: TS, total solids; DO, dissolved oxygen; BOD, biological oxygen demand; USPH, United States Public Health Service; ISI, Indian Standard Ins; WHO, World Health Organization; ICMR, Indian Council of Medical Research; –, not found.

Index